UNIFIED METHODS FOR VLSI SIMULATION
AND TEST GENERATION

THE KLUWER INTERNATIONAL SERIES IN ENGINEERING AND COMPUTER SCIENCE

VLSI, COMPUTER ARCHITECTURE AND DIGITAL SIGNAL PROCESSING

Consulting Editor
Jonathan Allen

Other books in the series:

Logic Minimization Algorithms for VLSI Synthesis. R.K. Brayton, G.D. Hachtel, C.T. McMullen, and A.L. Sangiovanni-Vincentelli. ISBN 0-89838-164-9.
Adaptive Filters: Structures, Algorithms, and Applications. M.L. Honig and D.G. Messerschmitt. ISBN 0-89838-163-0.
Introduction to VLSI Silicon Devices: Physics, Technology and Characterization. B. El-Kareh and R.J. Bombard. ISBN 0-89838-210-6.
Latchup in CMOS Technology: The Problem and Its Cure. R.R. Troutman. ISBN 0-89838-215-7.
Digital CMOS Circuit Design. M. Annaratone. ISBN 0-89838-224-6.
The Bounding Approach to VLSI Circuit Simulation. C.A. Zukowski. ISBN 0-89838-176-2.
Multi-Level Simulation for VLSI Design. D.D. Hill and D.R. Coelho. ISBN 0-89838-184-3.
Relaxation Techniques for the Simulation of VLSI Circuits. J. White and A. Sangiovanni-Vincentelli. ISBN 0-89838-186-X.
VLSI CAD Tools and Applications. W. Fichtner and M. Morf, editors. ISBN 0-89838-193-2.
A VLSI Architecture for Concurrent Data Structures. W.J. Dally. ISBN 0-89838-235-1.
Yield Simulation for Integrated Circuits. D.M.H. Walker. ISBN 0-89838-244-0.
VLSI Specification, Verification and Synthesis. G. Birtwistle and P.A. Subrahmanyam. ISBN 0-89838-246-7.
Fundamentals of Computer-Aided Circuit Simulation. W.J. McCalla. ISBN 0-89838-248-3.
Serial Data Computation. S.G. Smith and P.B. Denyer. ISBN 0-89838-253-X.
Phonological Parsing in Speech Recognition. K.W. Church. ISBN 0-89838-250-5.
Simulated Annealing for VLSI Design. D.F. Wong, H.W. Leong, and C.L. Liu. ISBN 0-89838-256-4.
Polycrystalline Silicon for Integrated Circuit Applications. T. Kamins. ISBN 0-89838-259-9.
FET Modeling for Circuit Simulation. D. Divekar. ISBN 0-89838-264-5.
VLSI Placement and Global Routing Using Simulated Annealing. C. Sechen. ISBN 0-89838-281-5.
Adaptive Filters and Equalisers. B. Mulgrew, C.F.N. Cowan. ISBN 0-89838-285-8.
Computer-Aided Design and VLSI Device Development, Second Edition. K.M. Cham, S-Y. Oh, J.L. Moll, K. Lee, P. Vande Voorde, D. Chin. ISBN: 0-89838-277-7.
Automatic Speech Recognition. K-F. Lee. ISBN 0-89838-296-3.
Speech Time-Frequency Representations. M.D. Riley. ISBN 0-89838-298-X
A Systolic Array Optimizing Compiler. M.S. Lam. ISBN: 0-89838-300-5.
Algorithms and Techniques for VLSI Layout Synthesis. D. Hill, D. Shugard, J. Fishburn, K. Keutzer. ISBN: 0-89838-301-3.
Switch-Level Timing Simulation of MOS VLSI Circuits. V.B. Rao, D.V. Overhauser, T.N. Trick, I.N. Hajj. ISBN 0-89838-302-1
VLSI for Artificial Intelligence. J.G. Delgado-Frias, W.R. Moore (Editors). ISBN 0-7923-9000-8.
Wafer Level Integrated Systems: Implementation Issues. S.K. Tewksbury. ISBN 0-7923-9006-7
The Annealing Algorithm. R.H.J.M. Otten & L.P.P.P. van Ginneken. ISBN 0-7923-9022-9.

UNIFIED METHODS FOR VLSI SIMULATION AND TEST GENERATION

by

Kwang-Ting Cheng
and
Vishwani D. Agrawal

AT&T Bell Laboratories
Murray Hill, New Jersey

Kluwer Academic Publishers
Boston/Dordrecht/London

Distributors for North America:
Kluwer Academic Publishers
101 Philip Drive
Assinippi Park
Norwell, Massachusetts 02061 USA

Distributors for all other countries:
Kluwer Academic Publishers Group
Distribution Centre
Post Office Box 322
3300 AH Dordrecht, THE NETHERLANDS

Library of Congress Cataloging-in-Publication Data

Cheng, Kwang-Ting, 1961-
Unified methods for VLSI simulation and test generation by / Kwang
-Ting Cheng and Vishwani D. Agrawal.
p. cm. — (The Kluwer international series in engineering and
computer science : #73)
Bibliography: p.
Includes index.
ISBN 0-7923-9025-3
1. Integrated circuits—Very large scale integration—Design and
construction—Data processing. 2. Computer-aided design.
3. Integrated circuits—Very large scale integration—Testing.
4. Integrated circuits—Very large scale integration—Computer
simulation. I. Agrawal, Vishwani D., 1943- II. Title.
III. Series: Kluwer international series in engineering and computer
science : SECS 73.
TK7874.C525 1989
621.39'5—dc20 89-34073
CIP

Printed in the United States of America

To Our Parents

And

To Our Wives

Table of Contents

PREFACE

This monograph reports our recent work on simulation-based methods for test generation. We have written it for CAD engineers, VLSI designers, test engineers, and researchers.

Most people who deal with digital circuits, realize that test generation for sequential circuits is a very difficult problem. The known algorithms, when programmed, have proved to be rather inefficient and computationally expensive. In the winter of 1986, we set out to look for a new solution. We noticed that simulators and test generators manipulate the same circuit description but use distinctly different algorithms. Simulators analyze logical behavior and delays of circuit elements while test generators only analyze the logical behavior. However, the high complexity of test generators makes it impractical to add any timing considerations in them.

We decided to develop a test generator using the principles of logic simulation. It would have two advantages: (1) It will deal with sequential logic with the same ease as a logic simulator, and (2) It can be easily incorporated in any existing logic simulator. We believe we accomplished both objectives.

The main idea in our method is to use a cost function. This cost function, which is computed from the logic simulation results, tells us how good an input vector is for testing a fault. Test generation now consists of modifying the input vector to reduce the cost.

To make the book complete, we have included tutorial material on logic simulation and test generation in Chapters 2 and 3, respectively. We have also included an extensive bibliography at the end. In Chapter 4, we formulate the principle of directed search and the role of cost function. In Chapter 5, we give a new form of simulation model known as the threshold-value model. This model adds controllability information to the normal Boolean gate model and thereby provides a simple way of computing the cost function. In Chapter 6, we develop a simulation-based test generator, *TVSET*, using the threshold-value model. While the threshold-value model provides a good understanding of the new test generation procedure, it is not the only way. In Chapter 7, we apply our method to a popular VLSI CAD tool, the concurrent fault simulator. The resulting test generator, *CONTEST*, uses the simple Boolean gate model and produces complete tests (including initialization vectors) for any general sequential (synchronous or asynchronous) circuit without human intervention.

The simplicity of the cost-function methods and the early success of our programs lead us to believe that such methods have a high potential in VLSI CAD.

Acknowledgment – We gratefully acknowledge the collaboration of Prathima

Agrawal in the development of *CONTEST*. We also acknowledge the help of Prof. S. C. Seth with the material in Chapter 2. We are thankful to Prof. Ernest Kuh of University of California at Berkeley for his support of the research reported in this book. When this work was done, one of the authors was a Ph.D. candidate at Berkeley and derived financial support from the NSF grant ECS-85-06901 and the State of California Microelectronics and Computer Research Opportunities Program. The collaboration between the two authors was made possible by a consulting support from the AT&T Bell Laboratories.

Chapter 1
INTRODUCTION

Even though testing of digital circuits is an established area of research, it requires revamping to keep pace with the rapid advances in integrated circuit technology. As the number of components on a chip and the gate to pin ratio increase, the testing problem becomes more difficult and the testing cost contributes to an increasingly larger proportion of the total product cost. Also, classical testing approaches would become obsolete and novel and revolutionary approaches will be required to handle the ever-increasing complexity of VLSI devices.

The primary task of testing is to detect or discover the physical defects produced during manufacturing processes. In general, logic circuits are tested by a sequence of input stimuli, known as *test vectors*, that check for possible faults in the circuit by producing observable faulty response at primary outputs. In this process, the most difficult part is the generation of a set of test vectors that will check faults at all points in the circuit and, therefore, uncover close to 100% of all possible defects in the chip.

1.1. THE TESTING PROBLEM

Fault Modeling. During chip fabrication, many types of defects can occur. For example, breaks in signal lines, lines shorted to ground, excessive delays, etc. In general, the effect of a fault is represented by means of a model. The most common model is the *single stuck-at fault.* In this model, we assume that any one signal in the circuit may have a fault such that this signal is fixed to either a logic 1 or a logic 0 irrespective of the input vectors. The single stuck-at fault model has been found to be effective in representing the behavior of faulty circuits, i.e., the test vectors that can detect a large number of single stuck-at faults can also detect most of the defects in the actual device. This model is simple to analyze because the faults are considered at the logic level which is independent of technology. Furthermore, the total number of modeled faults is at most twice the total number of signals in the circuit. This number can be further reduced by *fault collapsing* (see Chapter 2). We discuss more details of fault modeling in Chapter 2.

Test Generation. The problem of generating a test for a given fault has been proved to be NP-complete even for combinational circuits [1,2]. Although some test generation methods guarantee a test if one exists for a fault in a combinational circuit, the NP-completeness property necessitates the use of clever heuristics in practice. Various test generation approaches are discussed in Chapter 3. While test generation for purely combinational circuits is challenging due to the high circuit complexity, sequential circuits pose additional complications due to the presence of memory states.

Measures of Test Quality. The quality of tests is measured in terms of the size (or length) of the test sequence and the fault coverage. The length of the test sequence determines the time required to test the actual device on an automatic test equipment. The fault coverage is defined as the fraction of modeled faults detected by the test sequence. It is evaluated by fault simulation that computes the test response of the fault-free and the faulty circuits. From the simulation results, we can determine which faults are detected. It is desirable to achieve a high fault coverage. Almost 100% coverage is required for military applications. Even for commercial application, this percentage must be in high 90s. We will elaborate on this topic in Chapter 2.

1.2. THE TESTING TECHNIQUES

Large scale integration has added enormous complexity to the process of testing. This complexity is due mainly to the reduction in the ratio of externally accessible points (primary inputs and outputs) to internal inaccessible points in the circuit. We can classify the integrated circuit testing techniques into three broad categories, namely; (1) testing of purely combinational circuits or synchronous sequential circuits using scan type of design for testability (DFT) techniques; (2) self-testing circuits that generate their own test vectors using built-in hardware; and (3) testing of general digital (sequential) circuits with test vectors that are externally generated and applied.

For purely combinational circuits, a number of methods [3,4,5] are known that automatically generate tests with satisfactory fault coverage. For synchronous sequential circuits, scan design is often used to reduce the test generation problem to a combinational circuit testing problem that is considerably less difficult. In scan design [6], all memory elements of the circuit are chained into one or more shift registers such that a synchronous sequential circuit can be partitioned into a purely combinational circuit and a shift register that can be tested separately. While this technique has been successfully used in commercial systems, the 10-20% logic added for testability has performance and cost penalties that are not always acceptable.

In the Built-In Self-Test (BIST) approach [7], the circuit is designed to have a self-test mode in which it generates test vectors and analyzes the response. The objective is to apply all possible vectors to the combinational part of the circuit. In very large circuits, either the combinational portion is partitioned into independent sections or the whole circuit is tested by random vectors. However, the hardware overhead of BIST is even higher than the scan design; 20-50% test logic may have to be added for BIST.

Fig. 1.1. shows the DFT methodology of VLSI chip realization. Because of the hardware overhead penalty, a majority of VLSI chips manufactured today contain neither scan nor built-in self-test capability. Most of these chips contain sequential logic. The test generation problem for general sequential circuits is recognized to be very difficult, tedious, and as yet, unsolved. The memory elements contribute to the poor controllability and observability of logic. Most test generators for sequential circuits can perform reasonably well only on circuits with up to 1,000 gates. Therefore, VLSI designers manually develop test vectors using the knowledge of the functional behavior of the circuit. These tests are generated to exercise critical paths and functions with selected data patterns. In spite of the enormous effort, the quality of manually generated test vectors, as often verified by fault simulation, appears questionable.

There exists a critical need to develop an automatic sequential circuit test generator that can handle VLSI chips and, at a reasonable computing cost, achieve high fault coverage. This is the main objective of the present work. Equally important is the objective to combine the two processes of test generation and fault

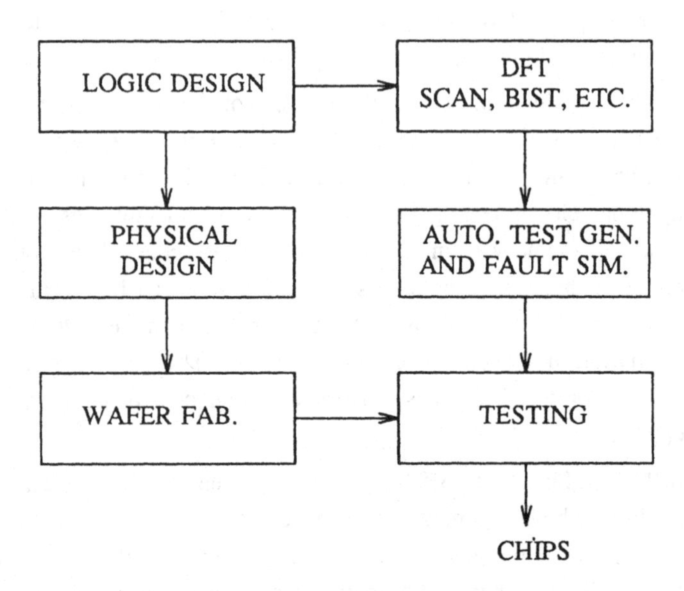

Fig. 1.1 Conventional methodology of VLSI chip realization with automatic test generation.

simulation. Ideally, we would like to simplify the chip design methodology as shown in Fig. 1.2. In this methodology, the hardware-expensive DFT is eliminated, yet the advantage of automatic test generation is maintained.

1.3. ORGANIZATION OF THE BOOK

This book presents an entirly new approach to test generation for general digital circuits. Using this approach, two algorithms are developed, namely, TVSET [8,9] and CONTEST [10,11]. TVSET uses a new modeling and simulation technique called the threshold-value simulation [12]. CONTEST extends the well known concurrent fault simulation method for test generation.

Chapter 2 introduces the concepts of logic modeling, event-driven simulation, fault modeling and fault simulation. Chapter 3 describes several test generation approaches that are respresentative of the previous work in this area. Specific situations responsible for the high complexity and poor performance of these approaches are also illustrated. Chapter 4 gives an overview of the new simulation-based method. Crucial to the development of TVSET is the threshold-

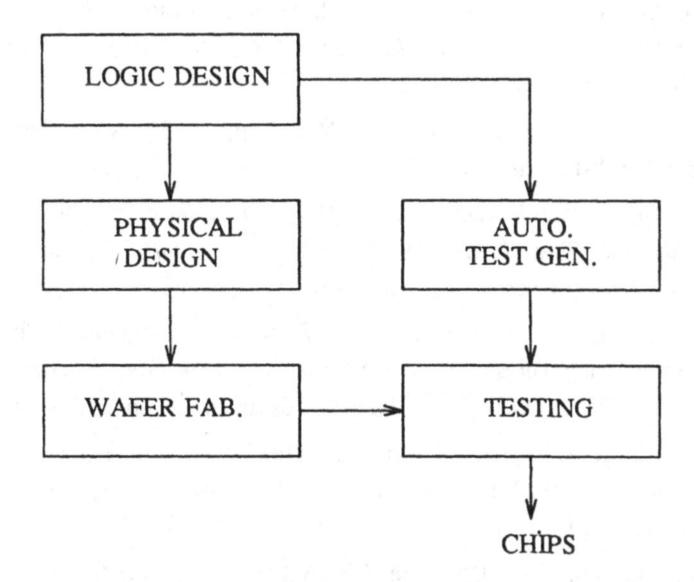

Fig. 1.2 Large VLSI realization with the proposed method.

value simulation technique discussed in Chapter 5. Chapters 6 and 7 describe the two algorithms, give their detailed implementation, and provide experimental results. Conclusions and future research directions are addressed in Chapter 8. Chapter 9 contains a detailed bibliography of the literature on test generation.

REFERENCES

[1] O. H. Ibarra and S. K. Sahni, "Polynomially Complete Fault Detection Problems," *IEEE Trans. Comp.*, Vol. C-24, pp. 242-249, March 1975.

[2] H. Fujiwara and S. Toida, "The Complexity of Fault Detection Problems for Combinational Logic Ciruits," *IEEE Trans. Comp.*, Vol. C-31, pp. 555-560, June 1982.

[3] J. P. Roth, W. G. Bouricius, and P. R. Schneider, "Programmed Algorithms to Compute Tests and to Detect and Distinguish Between Failures in Logic Circuits," *IEEE Trans. on Electronic Computers*, Vol. EC-16, pp. 567-580, October 1967.

[4] P. Goel, "An Implicit Enumeration Algorithm to Generate Tests for Combinational Logic Circuits," *IEEE Trans. Comp.*, Vol. C-30, pp. 215-222, March 1981.

[5] H. Fujiwara and T. Shimono, "On the Acceleration of Test Generation Algorithms," *IEEE Trans. Comp.*, Vol. C-32, pp. 1137-1144, Dec. 1983.

[6] V. D. Agrawal, S. K. Jain, and D. M. Singer, "Automation in Design for Testability," *Proc. Custom Integrated Circuits Conf.*, Rochester, NY, pp. 159-163, May 1984.

[7] *IEEE Design & Test of Computers*, Vol. 2, pp. 29-36, April 1985, Special issue on Built-In Self-Test.

[8] K. T. Cheng and V. D. Agrawal, "A Simulation-Based Directed-Search Method for Test Generation," *Proc. Int. Conf. Computer Design. (ICCD'87)*, Port Chester, NY, pp. 48-51, October 1987.

[9] K. T. Cheng, V. D. Agrawal, and E. S. Kuh, "A Sequential Circuit Test Generator Using Threshold-Value Simulation," *18th Fault-Tolerant Computing Symp. (FTCS-18) Digest of Papers*, Tokyo, Japan, pp. 24-29, June 1988.

[10] K. T. Cheng, V. D. Agrawal, and P. Agrawal, "Use of a Concurrent Fault Simulator for Test Vector Generation," *Proc. AT&T Conference on Electronic Testing*, Princeton, NJ, pp. 23-28, October 1987.

[11] V. D. Agrawal, K. T. Cheng, and P. Agrawal, "CONTEST: A Concurrent Test Generator for Sequential Circuits," *25th Design Automation Conf.*, Anaheim, CA, pp. 84-89, June 1988.

[12] V. D. Agrawal and K. T. Cheng, "Threshold-Value Simulation and Test Generation," *Testing & Diagnosis of VLSI & ULSI (Proc. NATO Adv. Study Inst. Como, Italy, June 1987)*, M. Sami and F. Lombardi, Editor, Kluwer Academic Publishers, Dordrecht, The Netherlands, 1988.

Chapter 2

LOGIC SIMULATION AND FAULT ANALYSIS

Knowing the response of every component in a complex design does not imply that the entire system will function correctly. The system response is usually obtained through simulation. Since our objective in this book is to present an unified methodology for simulation and test, an understanding of the concepts described in this chapter is essential. More specifically, we introduce the concepts of logic modeling, event-driven simulation, fault modeling and fault simulation. Readers who are familiar with these concepts, may need to take only a cursory look at the material. Those desiring more details are referred to a recent publication [1] from which this chapter is derived.

2.1. SIMULATION

Simulation normally refers to obtaining the response of a system from a model. For electronic systems both hardware and software models are used. While hardware models or *breadboards* are the traditional way of verifying designs, software models are becoming more popular due to their economy and accuracy derived largely from the advances in computing.

2.1.1. Circuit Modeling

Digital circuits are modeled as interconnections of functional elements. The interconnections are described using a hardware description language. Generally, the *level* of modeling refers to the amount of functionality that is included in the elements. At the highest level, commonly known as the *behavior-level*, the elements are large functional blocks described in programming languages like C or Pascal. The next lower level is the *gate-level* which consists of Boolean gates like AND, NAND, NOR, NOT and OR. Next is the MOS* transistor level also referred to as the *switch-level*. The lowest level is the *circuit-level* where components like transistors, resistors, capacitors, etc., are described through their electrical characteristics.

In order to effectively deal with the complexity of very large scale integration (VLSI), *mixed-level* models are used. In a large system, blocks of preverified designs may be modeled at the behavior level. Other portions of the same system may be modeled at the gate or transistor levels.

Another useful concept in describing VLSI systems is *hierarchy*. In a digital system, functional blocks like registers, adders, and multiplexers may be described as interconnections of logic gates. The system then can be described as an interconnection of these functional blocks. If the system uses a functional block repeatedly then the details of this block are described only once in the hierarchical description.

2.1.2. Signals

In a digital system, the structure of the circuit is not the only thing that is modeled. The signals flowing through interconnections must also be modeled. Real signals in an electronic system are voltages and currents. Strictly speaking, these are analog quantities. However, for digital systems, the common methodology is to model them as having discrete values. Most simulation systems use either three values (0, 1, X) or four values (0, 1, X, Z). The first two values, 0 and 1, are the *false* and *true* logic states. The third value, X, is used to represent the *unknown* or ambiguous state of a signal and the fourth state, Z, represents the state of a *floating* node in the circuit. Based on the specific situation, a floating node may or may not retain its previous value.

---

* MOS refers to the Metal Oxide Semiconductor technology which derives its name from the type of semiconductor devices used on chips [2]. Two types of devices are common. These are called the n-channel (NMOS) and the p-channel (PMOS) transistors. A popular integrated circuit technology, known as Complementary MOS (CMOS), employs both types of transistors.

Another essential attribute of signals is *time*. All signals are represented as functions of time. More appropriately, they can be considered as waveforms.

2.1.3. Modeling of Delays

All circuit elements manipulate the signals supplied to their input port and produce the resulting signals at their output port. The manipulation of signals, however, takes finite time. Thus, irrespective of their speed, all electronic circuits involve delays. The delay is modeled in many different ways.

The simplest method is to assume that the elements (functional blocks, logic gates or transistor switches) have *zero delay*. This assumption works well if the interconnections involve no feedback paths. The zero-delay model can be effectively used to analyze combinational circuits that have no memory states.

When the circuit contains feedback paths or memory elements (e.g., flip-flops) it is necessary to maintain the order in which signals change. Under fairly general conditions, it is possible to maintain the proper sequence of events (signal changes) by a *unit-delay* model. In this model, each element is assumed to have one unit of delay. Since actual delays of elements are not all equal, the actual interval of time that this unit represents is meaningless. The unit-delay models is generally used for logic verification of gate-level and switch-level circuits. This model is also very popular for fault simulation discussed in a later section.

While the unit-delay model can verify the logical behavior of a digital circuit, it is inadequate for analyzing the timing behavior. For the timing behavior, a *multiple-delay* model is used. Each element is assigned a delay which is an integer multiple of a time unit. The time unit can be 1 nanosecond, 1 picosecond, or some such interval. Sometimes separate rise and fall delays are specified. It is also possible to specify propagation delays for interconnecting paths.

Most digital circuits can be analyzed by an appropriate discrete delay model as mentioned above. There can be situations, however, where a continuous delay model is necessary. Certain interconnections of bidirectional MOS devices (e.g., bus structures) and mixed analog-digital circuits are some examples requiring a fine-grain timing analysis. For VLSI circuits, mixed-mode analysis [3], where different components of the circuit are modeled at the appropriate level of timing accuracy, is perhaps the right way.

2.1.4. Simulator

A simulator is a computer program whose inputs are the circuit description and the primary input signal description or the *stimuli*. The stimuli can be specified as waveforms of the input signals. However, for digital circuits, a popular way is to specify them as *vectors*. A vector contains the values of all input signals. Whenever, one or more inputs change, a new vector must be specified. In most circuits, input signal changes are synchronized with some periodically changing clock signal. Thus, the input stimuli are a sequence of vectors applied at specified periodic intervals.

The simulator computes the response of the circuit. Earliest simulators were *compiled-code* simulators. Entire circuit description was converted into the form of a computer program. Execution of this program with the stimuli as data would then lead to the circuit response. There are, however, difficulties with this methodology. Since signal changes propagate from inputs to outputs, the circuit is levelized from inputs to outputs and the components are evaluated accordingly. A problem arises in the case of a feedback. Levelization is now impossible and a compiled-code simulator could give incorrect result. Circuits with components having different delays are also difficult to simulate.

It has been recognized that in a digital circuit, typically, at any instant only about 10% or fewer gates are active. Larger circuits tend to have smaller activity. However, the location of this activity changes with time. A compiled-code simulator takes no advantage of the low activity and evaluates all elements.

Accurate and efficient simulation of delays, feedbacks, and the dynamically changing activity is possible by the event-driven method described next.

2.1.5. Event-Driven Simulation

In the simulation terminology, signal changes are called the *events*. An event is characterized by the signal name, the type of change, and the time of change. A signal is assumed to retain its value until its source (e.g., the gate generating the signal) produces an event. A gate, on the other hand, can not produce an event at its output, unless some event causes a change at its input. An event-driven simulator, therefore, simply follows the paths of events in the circuit. The simulator also deals with the problem of simultaneous events by analyzing all events occurring at a time before analyzing any events that would take place in the future.

The general process of event-driven simulation can be easily understood from Fig. 2.1. The time, over which the circuit activity is analyzed, is divided into some suitably selected unit. All delays in the circuit are specified in terms of this time unit. The simulator contains a circular stack known as the *time wheel* [4]. Each element of this stack, called the *time slot*, represents one unit of time. The number of time slots in the time wheel should not be less than the largest delay of any element in the circuit.

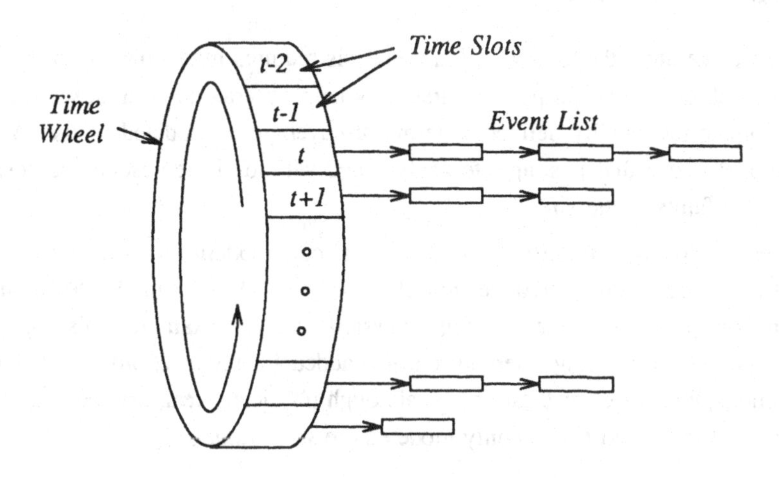

Figure 2.1 Event-scheduling in simulation.

Each time slot contains a pointer to a list of events that are to take place at the corresponding time. Since the number of events at any time can vary, the events are usually stored in a linked list. Primary input changes also produce events on the input signals. As explained above, an event is described by specifying the signal name, the new value, and the time of change. The signal name and new value are stored in the event list while the time of change is set by attaching the event to the appropriate time slot.

The simulator processes all events attached to the *current time slot*. Suppose, we denote the current time as t. Processing of an event means the following: (1) change the signal value, (2) evaluate all gates (or blocks) that now become potentially active due to an input signal change (this is done by tracing the fanout of the signal changed by the event), (3) if the output of any potentially active gate changes, then schedule the corresponding event. For scheduling an event, the simulator considers the delay of the gate whose output is changing. If the delay of that gate is d units, then the new event is attached to the event list of the time slot representing $t+d$.

Once an event is processed, i.e., all potentially active gates are evaluated and all new events are scheduled, the parent event is removed from its event list. When the event list of the time slot t becomes empty, the current time is advanced to $t+1$ and further event processing continues.

2.2. FAULT MODELING

Just like any other analysis, fault analysis requires modeling (or abstraction). Fault models serve two purposes. First, they help generate tests, and, second, they help evaluate test quality defined in terms of coverage of modeled faults. A good fault model is one that is simple to analyze and yet closely represents the behavior of physical faults in the circuit.

Fault modeling is strongly related to circuit modeling. Consider a digital circuit described as an interconnection of logic gates. Even a moderate amount of imagination provides numerous fault possibilities, for example, missing gates, wrong gate types, missing interconnections, added interconnections, shorted interconnections, etc. Most of these faults, although physically real, are too complex to model. The faults most commonly modeled are *stuck faults*.

2.2.1. Stuck Faults

Stuck faults are not only the simplest faults to analyze, but they also have proved to be very effective in representing the faulty behavior of actual devices. The simplicity of stuck faults is derived from their logical behavior; these faults are often referred to as *logical faults*. One of the earliest discussions on stuck faults was given by Poage [5].

Stuck faults are assumed to affect only the interconnections. Possible fault sites are the inputs and outputs of gates. Each line can have two types of stuck faults: stuck-at-1 and stuck-at-0. Thus, a line with a stuck-at-1 fault will always have a logical value 1 irrespective of the correct logical output of the gate driving it.

In general, several stuck faults can be simultaneously present in a circuit. A circuit with n lines can have 3^n-1 possible stuck line combinations. This is because each line can be in any one of the three states: stuck-at-1, stuck-at-0, or fault-free. All combinations except the one with all lines in the fault-free state are counted as faults. It is easy to realize that even with moderate values of n, the number of multiple stuck faults will be very large; therefore, in practice, we only analyze *single* stuck faults. An n-line circuit will have $2n$ single stuck faults, a number that can be further reduced by *fault collapsing*.

2.2.2. Fault Collapsing

Two faults are called *equivalent* if their effect is indistinguishable at the outputs of a circuit, which means that any test detecting one of them will also detect the other. Selecting one representative fault from each class of equivalent faults is called *equivalence fault collapsing*. Computationally, this is an intractable problem in its general form. In practice, incomplete yet substantial fault collapsing may be possible with little computational effort. A good example is the collapsing of faults associated with the inputs and the output of a logic gate. Consider an n-input AND gate with none of the inputs directly observable. It is easy to see that an input stuck-at-0 (it is a common practice to write *a stuck-at-i* as *a s-a-i*) is equivalent to the output s-a-0; there is no way to distinguish between the two faults by observing the inputs and outputs of the circuit. For the purpose of test generation, therefore, we only consider $n+2$ of the $2n+2$ faults associated with the AND gate: s-a-1 on each input and s-a-1 and s-a-0 on the output. Similarly, in an n-input OR gate, we test for s-a-0 on each input and s-a-1 and s-a-0 on the output. Testing considerations for NAND and NOR gates are similar.

A three-value logic simulator is effective in finding equivalent faults that may be separated by several gates in a circuit. The three values used in simulation are 0, 1, and X ≡ *unknown or don't care*. Consider the circuit of Figure 2.2. A value 0 on line *a* uniquely (i.e., irrespective of the other inputs of the NAND gate) forces a 1 on line *e*, and, thereby, forces a 1 on line *h*. This is easily checked by setting *don't care* values on all other input lines and noticing that the values of lines *e* and *h* are still unique. Thus, the faults, *a* stuck-at-0, *e* stuck-at-1, and *h* stuck-at-1, are equivalent.

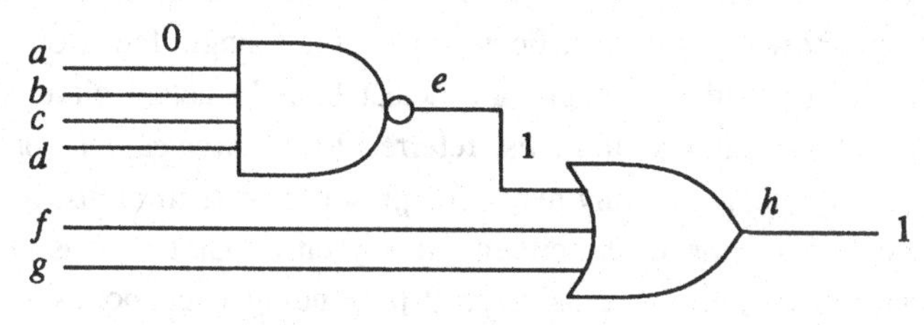

Figure 2.2 Collapsing of stuck faults.

The illustration above can be generalized. To find if a fault, say *a* s-a-*i* can be collapsed with another fault, we carry out three-value simulation after initializing line *a* to *i* and all other lines to *X*. If, as a result of simulation, another line, say *b*, is forced to a binary value *j*, then the fault *a* s-a-*i* is equivalent to *b* s-a-*j*.

This simple procedure has been used in an MOS simulator where the circuit is described as an interconnection of transistors [6].

In equivalence fault collapsing we only collapse the faults that are indistinguishable. If we are prepared to give up on diagnostic resolution (ability to distinguish between faults), more collapsing is possible. This is accomplished by using the concept of *fault dominance* as explained next. In very large scale integration (VLSI) circuits, where coverage of faults rather than their exact location is the overriding consideration, *dominance fault collapsing* may be desirable.

Consider two faults f_1 and f_2. Suppose all tests for f_1 also detect f_2 but only some of the tests for f_2 detect f_1. Then f_2 is said to *dominate* f_1. This definition of dominance, originally given by Poage [5], also appears in text books [7]. However, some authors have used the term dominance in an opposite sense [8]. If we had to pick one to detect, obviously, we are safer to take f_1. Even though, at times, it may be a little harder to find a test for f_1, this test is guaranteed to cover f_2.

In an AND gate, the output s-a-1 dominates any input s-a-1. Thus if we desire dominance fault collapsing, then for an n-input AND gate we need to consider only $n+1$ faults.

Equivalence and dominance fault collapsing may be used to reduce the number of faults that must be considered for detection. In the AND-gate example above, we found that the two stuck faults on the output line can be collapsed with appropriate input-line faults. This type of collapsing can be repeatedly used until one arrives at a *checkpoint* defined as either a primary input or a fan-out branch [7]. It has been shown that it is sufficient to consider single faults on checkpoints in a circuit as long as all such faults are detectable. In actual circuits, however, there can be a small number of faults, referred to as *undetectable* or *redundant* faults, that are not detected by any test. The presence of such faults, by definition, does not cause malfunction in the circuit. If any checkpoint fault is undetectable, then additional faults must be considered [9]. Finding checkpoints in sequential circuits requires further analysis [10].

In sequential circuits, fault collapsing is often accomplished through multiple passes. More details may be found in [11, 12]. In other related work, fault set reduction algorithms have been reported by Goel [13] and Cha [14].

2.2.3. Other Fault Models

Most VLSI designers encounter situations where stuck type fault model may not be suitable. Two common cases are the memory and programmable logic array (PLA) blocks. For memories, stuck fault model is still used for I/O registers and address decoders. However, for the storage cell array, the faults generally modeled are [15]: (1) single cell stuck at 1 or 0, (2) adjacent cell coupling, and (3) pattern-sensitive faults. Like memories, PLA fault models are also closely related to their structure. These include cross-point faults and bridging faults. For the purpose of logic and fault simulation, a PLA may be modeled as its two-level logic implementation. Even though stuck type faults in this implementation are not a true representation of PLA faults, they are easy to analyze and are widely used. Cross-point and bridging faults can be represented in the two-level logic model by adding extra gates [1].

While the above described fault models are used in today's design environment, there are others that are gaining importance due to the changing technology. Some of these are the CMOS stuck-open faults, functional faults in microprocessors, and delay faults [16].

2.3. FAULT SIMULATION

Simply stated, a perfect chip test must meet two criteria: reject every bad chip and pass every good one. Since the second requirement is not difficult to fulfill, we will focus on how well the first criterion can be met. Suppose a total of N chips passed the test and there are M bad chips among them. Then the fraction M/N, sometimes called the *reject ratio* of the tested product, is a measure of deficiency of the test. Unfortunately, appealing as it may be as an indicator of test quality, accurate reject-ratio data are hard to obtain in practice. Instead, an indirect but easier-to-estimate measure is used. It is called *fault coverage* and is defined as the percentage of modeled faults detected by the test. A good (or acceptable) test may be defined as one that achieves a certain minimum percentage of fault coverage (typically, in the high 90s). Accurate evaluation of fault coverage requires the use of a fault simulator, often working with the model of the circuit at just one level (e.g., transistor) [6], switch [17], or logic gate [18]. However, with increasing circuit densities, complete fault simulation using a low-level *non-hierarchical* circuit description is becoming very expensive and time consuming. Hierarchical [19] simulators are being developed in response to this need.

2.3.1. Methods of Fault Simulation

Given a set of faults and a set of input vectors, a fault simulator must find out which faults are detected by the input vectors. Production-quality fault simulators at the logic-gate level were available well before the advent of large scale integration (LSI) circuits. Recently, fault simulators at the transistor level have been developed for more accurate modeling of faults.

The simplest method of simulating faults is the *serial fault simulation*. A single fault is introduced in the circuit and simulation is run like the true-value simulation. Response is compared with the stored response of the fault-free circuit. As soon as the fault is detected, the current simulation is suspended and a new simulation is started with another fault. While this method has found some success in the hardware accelerator based simulators, in most commercial simulators, an effort is made to reduce computation time by simulating more than one fault in one pass for a given input vector. The simplest such technique is *parallel fault simulation* [20] in which a computer word of W bits is associated with each line in the circuit. During a pass through the simulator, each bit at a particular position would be associated with the circuit which has a specific single fault (or no fault). After simulation, the bit would store the value on the line in the associated circuit. Before the beginning of a pass, a set of $(W-1)$ as-yet-unsimulated faults is chosen, where W is the word size of the computer. The remaining bit is used to simulate the fault-free circuit. Gates whose inputs/outputs are directly affected by a selected fault are flagged. When a flagged gate is simulated, the effect of the fault is injected into appropriate bits of the words representing its inputs and outputs. Each fault simulation pass starts at the primary input lines and proceeds in a breadth-first fashion towards the primary outputs. For each gate, fault simulation amounts to computing the output word as the gate's logical function of its input words. At the end of the pass, the word at each primary output can be examined to determine which of the simulated faults would be detected at that output. This is done by comparing the response of each faulty circuit against the fault-free circuit. As logical operations on computers can be carried out at the word level, $W-1$ different faults are simulated in *parallel* in each pass; hence, the name of the method. A total of F faults would be simulated in $F/(W-1)$ passes.

The ultimate, in terms of reducing the number of passes, is the *deductive fault simulator*[21]. It needs just one pass for simulation independent of the number of faults simulated. The basic idea here is to associate with each line a list of just those faults sensitized to that line (that is, signals on the line for the normal and faulty circuits are different) by the simulated input vector. Simulation of a gate requires deducing the fault list at the gate output from the input fault lists. This is illustrated for a 2-input AND gate shown in the example of Figure 2.3. The gate is assumed to be embedded in a larger circuit. The fault lists associated with the inputs a and b of the gate are as shown; these would have been computed already in previous steps of the algorithm. The signal values shown are for the fault-free circuit. Consider the effect of the fault f_1, appearing in L_a but not in L_b, on the gate output. The signal value on line a will change from 1 to 0 but that on line b will remain unchanged. Thus, the output will stay at 0, so f_1 cannot be in the fault list L_c. Indeed, no fault in L_a can occur in L_c since the output would not change under the fault. Next, consider the fault f_8, which is in L_b but not in L_a. Such a fault will change the values on lines b and c and, hence, must be included in L_c. Additionally, the fault "c stuck at 1" (or c_1) must be in the output fault list since it complements the normal value. Thus we obtain the expression for the output fault list shown in the figure.

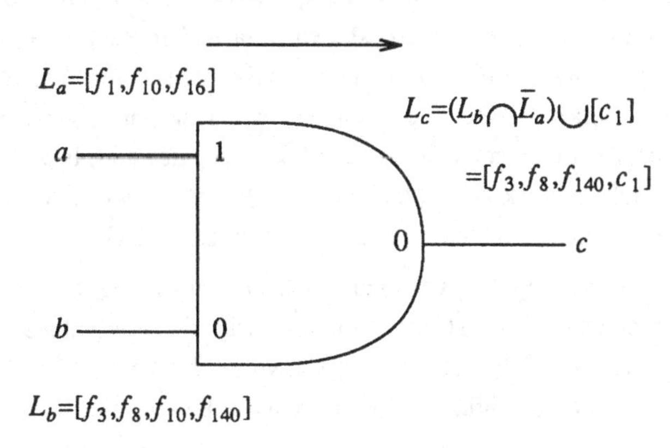

Figure 2.3 Fault lists in deductive simulation.

Compared with parallel fault simulation, the penalties paid for a single-pass in deductive simulation are (1) dynamically varying storage for fault lists associated with each line and (2) more complex processing of gates requiring set operations on the line fault lists.

We note that the output fault list computations are essentially dynamic: The expression for the output list is not just a function of the gate type, it also depends on the signal values on the gate inputs. Interestingly, the output fault list may change even though the gate inputs and outputs remain unchanged. This is because a line's fault list may change even when its value does not. For example, when the normal value on line a changes to 0, the value on line c is not changed, but L_c may change nevertheless. This "fault-list event" must be propagated through all gates to which line c is an input. This undesirable characteristic of the deductive method is absent from *concurrent simulation* [22] to which we shall turn next.

The basic idea behind concurrent fault simulation is quite simple. Typically a fault changes very few signal values in a large circuit. Thus, most, if not all, of the information for simulating a fault is contained in the "good-circuit" simulation. In concurrent simulation, the good circuit is simulated in its entirety, but a faulty one is simulated only for gates whose *states* differ from their good-circuit states. For logic gates, the state is simply the combination of input and output values, however, the concurrent method is general enough to handle arbitrary elements with stored states.

As an example, consider the two-gate circuit in Figure 2.4a with signals 1, 0, and 1 applied to the inputs a, b, and c, respectively. The figure shows the true value simulation for this vector. Also shown, attached to each gate, is a list of "faulty gates" whose states differ from their good-circuit states. These are identified by the faults marked inside. For example, when the fault "b stuck at 1" (b_1 in the figure) occurs, it changes the states of both the gates and, hence, appears on both lists. The fault "a stuck at 0" (denoted as a_0) does not cause the state of gate $G\,2$ to change and, hence, does not appear in the fault list of $G\,2$.

Next, suppose the input a was changed from 1 to 0 (Figure 2.4b). Since an input of $G\,1$ has changed, it must be resimulated with the new values. The resulting true values and the faulty gates are as shown in the figure. The new list for $G\,1$ was obtained from the original list through the following steps:

(1) The change in input a is reflected in the good as well as the faulty gates of $G\,1$, and the new output state of each gate is determined through simulation.

(2) If there are any faulty gates with state identical to the good-gate state, they are deleted from the list. In the example, this causes deletion of the entry for a_0.

(3) If there are any new faulty gates whose states are different from the good-gate state, they are added to the list. In the example, the entry for a_1 gets added because of this reason.

(4) Finally, if the output state of the good gate changed as a result of these steps, a good event is scheduled for the gates fed by the output of $G\,1$. Similarly, if the output state of a faulty gate changed as a result of the above steps, the corresponding faulty event is scheduled for the gates fed by the $G\,1$ output. In the example, no good event will be scheduled because the output in the good circuit has not changed. However, for the fault b_1 the

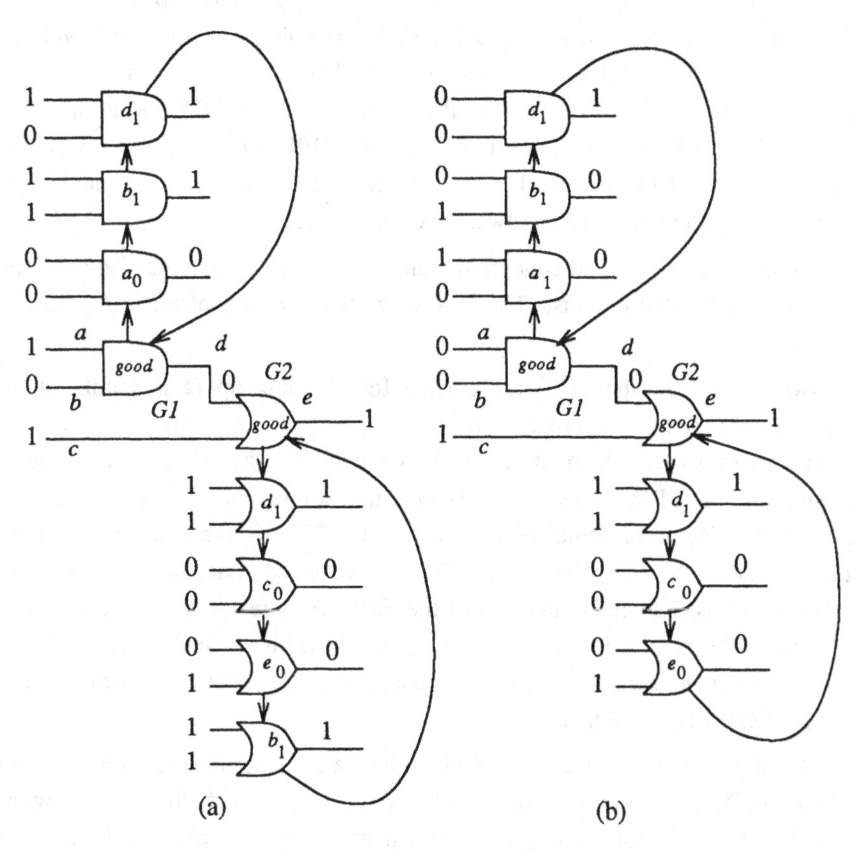

Figure 2.4 Concurrent fault simulation example.

output changes from 1 to 0, so the corresponding faulty event is scheduled at $G2$.

The faulty event at $G2$ is next processed. This involves a search for an entry for this event in the list of $G2$, changing it appropriately, and simulating the gate with the changed value. In this case, after simulation, the faulty-gate state coincides with the good-gate state; hence, the entry for b_1 is deleted from the list of $G2$. Note that the simulation did not cause a change in the output of the faulty gate so no further events need be scheduled.

The term *concurrent* is derived from the fact that each faulty gate carries enough information for independent simulation of the associated fault. This is considerably more information than just the fault index used in deductive simulation. Thus the speed is gained in concurrent simulation at the expense of additional dynamic storage per node in the circuit. There is more to the concurrent method, however, than indicated by this simple example. The ease with which it can be adapted to any level of simulation is evident in the following sample of implementations: FMOSSIM [17] at the switch level, MOTIS [6] at a mixed level, and the CHIEFS hierarchical fault simulator [19]. The ability of concurrent simulation to evaluate each faulty gate independently makes it the method of choice for implementations on hardware accelerators [23].

Parallel, deductive, and concurrent simulation may be used with both combinational and sequential circuits. The following methods, in contrast, apply only to combinational circuits.

Hong [24] describes a clever scheme for speeding up fault simulation by partitioning the circuit in terms of its *fanout free regions* (FFRs). An FFR is defined as a maximal subcircuit which does not include an internal fanout stem. The output of an FFR is either a primary output or a fanout stem. In Hong's algorithm the simulation of faults at the outputs of FFRs is done by a traditional method in the first step. Then each FFR is processed independently and efficiently by a process linear in the size of the FFR. As a result, the detectability of each fault in the FFR is known. Hong's approach has been further refined in the *critical path tracing* method [25] and in *fault_blaster* [26], both of which sacrifice the goal of exact fault simulation.

When *grading* (evaluating the fault coverage of) a set of test vectors by any of these methods, one would start with an initial list of faults containing all modeled faults. After simulating the first test vector against this list, the detected faults would be removed from the list before simulating the next vector. This *fault dropping* results in much faster evaluation of the cumulative fault coverage.

The methods described so far perform fault simulation one vector at a time to determine the fault coverage in a given list of faults. A recently proposed fault simulation method for combinational circuits, known as *parallel pattern single fault propagation (PPSFP)*, makes a departure from per-vector fault evaluation dictated by the traditional methods. It simulates 256 vectors in parallel and runs sequentially through the good and the faulty circuits to evaluate the fault coverage of these 256 vectors [27].

2.3.2. A Comparison of Fault Simulation Methods

The process of fault simulation can be likened to that of shooting at a progressively diminishing target. Initially, even a randomly chosen test vector is likely to detect many undetected faults, but as their number decreases this strategy produces diminishing results. Figure 2.5 shows a typical curve of undetected faults versus number of tests. It can be partitioned into two phases where random test generation is seen to be a productive strategy for Phase I (the exponential decay part of the curve). There is a gradual transition from Phase I to an almost linear decay in Phase II.

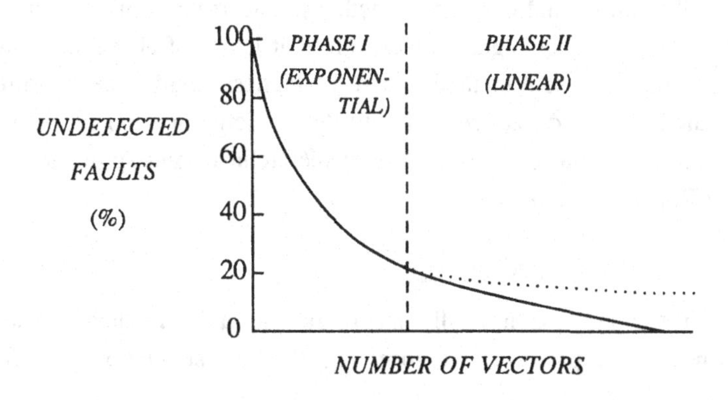

Figure 2.5 Undetected faults versus number of test vectors.

Empirical evidence suggests the transition point to lie somewhere between 35 and 15 percent of undetected faults. Taking such a curve as the model, it has been estimated [28] that the simulation costs for a network with G gates grow proportional to G^3 and G^2 for parallel and deductive (or concurrent) simulations, respectively. A concurrent fault simulator at the transistor level is reported to take 1527 seconds (approx. 25 minutes) of computation time for simulating a 9,478-transistor circuit on a 12-MIP computer [6]. A set of 2,241 vectors were used in this simulation. According to the quadratic rise in simulation time, a 100,000-transistor circuit would require over 47 hours on the same machine! Clearly, brute-force simulation of large circuits is not a practical proposition even for the currently achievable circuit densities. Attractive alternatives are to use statistical sampling techniques [29] or hardware accelerators.

A recent theoretical result sheds further light on the cost of fault simulation [30]. By relating the fault simulation of a specific class of combinational circuits to a well-known problem in the complexity theory (the Boolean matrix multiplication problem), it essentially rules out any hope for a linear time algorithm for per-vector fault simulation. It can be easily shown from elementary considerations that the worst-case time need not grow faster than a quadratic rate in circuit size. Thus the theoretical complexity appears to be between linear and quadratic. On the said class of circuits, the paper shows that all of the standard fault simulation methods (parallel, deductive, concurrent, and critical-path-tracing) would achieve the quadratic complexity.

Another way to compare fault simulation methods is to analyze their capabilities. One such comparison was carried out by Levendel and Menon [31] for the parallel, deductive, and concurrent methods. The metrics of comparison were abilities to handle multiple signal values, different levels of abstraction, and accurate timing. The concurrent method scored the highest on all three measures. The parallel method came a distant second with the deductive not too far behind it. A more recent analysis that justifies the use of hierarchical fault simulation has been reported by Rogers and Abraham [32].

2.4. FAULT SIMULATION TOOLS

VLSI CAD systems normally incorporate a fault simulator. These fault simulators have three parts: (1) preprocessing, (2) fault simulation, and (3) output analysis.

A fault simulator processes the circuit connectivity description in much the same way as a true-value simulator. Often, true-value simulation for design verification precedes fault simulation, and a preprocessed or compiled description of the circuit already exists. Additionally, the fault simulator makes a fault list of all stuck faults on the inputs and outputs of gates and the functional blocks. Some fault simulators provide an option to include transistor faults. If the circuit contains functional blocks (e.g., memories or other blocks described in some high-level language), then stuck faults are modeled on the inputs and outputs of those blocks. The fault list is processed to collapse equivalent faults. The user can select or remove any faults from the fault list or request a random sample of given size. User inputs also include specification of the signals to be monitored (e.g., all primary outputs) and the strobe positions. Many fault simulators allow user selectable options for handling of race and oscillation faults. These faults may introduce potentially unstable states in the circuit and are often represented by the unknown value (X) in simulators. If an unknown state due to a fault propagates to a visible point, the fault can be considered as potentially detectable.

The second part of a simulator performs its main function: simulation of the circuit with selected faults by using the given vectors. It keeps the detection data (time and output of detection) for each fault and also stores the true-value response. Most simulators drop a detected fault from consideration once it is detected. Race and oscillation faults may also be dropped at user option.

Since fault simulators use a large amount of computing resources, the normal practice is to divide the vector set according to the available memory in the automatic test equipment (ATE). Each vector set (often called a *vector load*) is run through the simulator by using the list of undetected faults up to that point. At the end of a run, the simulator can store the internal states of the circuit in a checkpoint data set for use by the next vector set. However, if the ATE requires each vector set to reinitialize the circuit, the simulator checkpointing will give a coverage that is too optimistic. Also, for realistic fault detection data, the user should set the observation strobes in the same way as the ATE's strobes.

The results of the second part are fault coverage and the expected response. The latter is provided to the test program. The third part of a fault simulator analyzes the fault coverage data and displays it in an easy-to-interpret form. A popular way of presentation is a graph showing fault coverage as a function of the vector number in the test sequence. The simplest way, of course, is to list faults under the following categories: detected, undetected, race, and oscillation. These lists may be used by other programs such as automatic test generators. Displaying undetected faults on the schematic can be a great help in manual test generation. For an undetected fault, tracing the activity caused by this fault can also be useful to the designer in writing a test.

Another effective method of displaying the fault simulator result would be to display the detected faults on the chip layout. Such a display will easily isolate the portions of the chip where fault coverage needs enhancement.

2.5. FAULT COVERAGE AND PRODUCT QUALITY

Independent of whether the whole class of single line stuck faults is simulated or only a sample thereof, the obtained fault coverage is at best an imperfect measure of test effectiveness. This must be so because a fault simulator can not evaluate the coverage of actual physical faults (shorts or opens in metal, diffusion, or polysilicon, shorts between layers, parametric irregularities, etc.). Multiple faults are also not simulated because of their very large number, even though with the small geometries used, a processing defect is quite likely to lead to multiple faults. The obtained value of fault coverage may also depend on the specific simulator used for evaluation since different simulators could employ different criteria to detect fault-induced races and oscillations. Redundant lines are not uncommon in real circuits though they may not always be identifiable because of the provable intractability of the problem. The effect of unidentified redundancies is an unduly pessimistic fault coverage.

In spite of the above reservations, fault coverage data provided by simulators continue to be relied on as a *figure of merit* for a test. One may well ask, how is this figure of merit related to the quality of the tested chips? A quantitative answer to this question can be given based on a model of fault distribution on a chip. In one such model, it is assumed that a random number of logical faults are spawned by each physical defect on the chip. The physical defects themselves are randomly distributed over the chip but form clusters for a variety of processing related reasons. Thus a compound distribution is used for logical faults [33,34] with an appropriate clustered distribution used for physical defects. It is possible to estimate the parameters of the compound distribution from the wafer-level test data and produce the curve shown in Figure 2.6. The vertical ordinate in the figure is *reject ratio* defined in Section 2.3. The horizontal axis represents the fault coverage for single stuck-type faults.

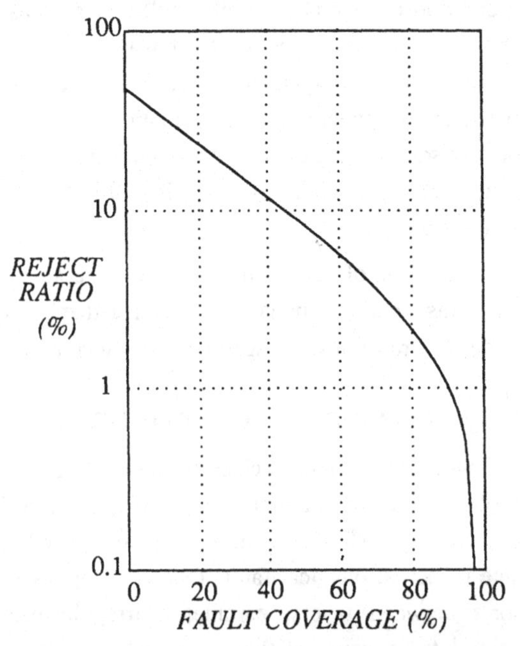

Figure 2.6 Reject ratio versus fault coverage.

Interestingly, it is observed that, making a transition to a finer feature size, while keeping other parameters fixed, has the effect of moving the curve to the left. This means that for denser chips actually a lower value of fault coverage would suffice for a given value of reject ratio; an encouraging result considering the disproportionately high costs involved in increasing the fault coverage in Phase II of Figure 2.5 to cover such faults.

In a related work by Williams and Brown [35] a uniformly random defect distribution model is used for logical faults. They use the term *defect level* to denote what we have called as reject ratio. Although the results have been applied to both chips and boards, the uniformly random fault distribution gives pessimistic results for chips where the defects may be clustered.

REFERENCES

[1] V. D. Agrawal and S. C. Seth, *Tutorial – Test Generation for VLSI Chips,* IEEE Computer Society Press, Washington, D.C., 1988.

[2] N. G. Einspruch, *VLSI Handbook,* Academic Press, Orlando, FL, 1985.

[3] V. D. Agrawal, A. K. Bose, P. Kozak, H. N. Nham, and E. Pacas-Skewes, "Mixed-Mode Simulation in the Motis System," *Journal of Digital Systems,* Vol. V, pp. 383-400, 1981.

[4] E. G. Ulrich, "Exclusive Simulation of Activity in Digital Networks," *Communications of the ACM,* Vol. 12, pp. 102-110, February 1969.

[5] J. F. Poage, "Derivation of Optimum Tests to detect Faults in Combinational Circuits," pp. 483-528 in *Proc. Symp. on Mathematical Theory of Automata (April 1962),* Polytechnic Press, New York, 1963.

[6] C. Y. Lo, H. N. Nham, and A. K. Bose, "Algorithms for an Advanced Fault Simulation System in MOTIS," *IEEE Trans. CAD,* Vol. CAD-6, pp. 232-240, March 1987.

[7] M. A. Breuer and A. D. Friedman, *Diagnosis & Reliable Design of Digital Systems,* Computer Science Press, Rockville, MD, 1976.

[8] J. A. Abraham, "Fault Modeling in VLSI," in *VLSI Testing,* ed. T. W. Williams, North-Holland, Amsterdam, The Netherlands, 1986.

[9] M. Abramovici, P. R. Menon, and D. T. Miller, "Checkpoint Faults Are Not Sufficient Target Faults for Test Generation," *IEEE Trans. on Computers,* Vol. C-35, pp. 769-771, August 1986.

[10] S. J. Chang and M. A. Breuer, "A Fault-Collapsing Analysis in Sequential Logic Networks," *Bell Syst. Tech. Jour.,* Vol. 60, pp. 2259-2271, Nov. 1981.

[11] E. J. McCluskey and F. W. Clegg, "Fault Equivalence in Combinational Logic Networks," *IEEE Trans. Computers,* Vol. C-20, pp. 1286-1293, November 1971.

[12] D. R. Schertz and G. Metze, "A New Representation for Faults in Combinational Digital Circuits," *IEEE Trans. Computers,* Vol. C-21, pp. 858-866, August 1972.

[13] P. Goel, "The Feed Forward Logic Model in the Testing of Large Scale Integrated Logic Circuits," *Ph.D. Dissertation*, Carnegie-Mellon University, Pittsburgh, PA, September 1973.

[14] C. W. Cha, "Multiple Fault Diagnosis in Combinational Networks," *Proc. 16th Des. Auto. Conf.*, San Diego, CA, pp. 149-155, June 1979.

[15] M. S. Abadir and H. K. Reghbati, "Functional Testing of Semiconductor Random Access Memories," *ACM Computing Surveys*, Vol. 15, pp. 175-198, September 1983.

[16] V. D. Agrawal and S. M. Reddy, "Fault Modeling and Test Generation," in *VLSI Handbook*, ed. J. DiGiacomo, McGraw-Hill, New York, 1989. Chapter 8

[17] M. D. Schuster and R. E. Bryant, "Concurrent Fault Simulation of MOS Digital Circuits," *Proc. Conf. Adv. Res. in VLSI*, Cambridge, MA, pp. 129-138, January 1984.

[18] M. Abramovici, J. J. Kulikowski, P. R. Menon, and D. T. Miller, "SMART and FAST: Test Generation for VLSI Scan-Design Circuits," *IEEE Design & Test of Computers*, Vol. 3, pp. 43-54, Aug., 1986, Also *Proc. Int. Test Conf.*, Philadelphia, PA, Nov., 1985, pp. 45-56.

[19] W. A. Rogers and J. A. Abraham, "CHIEFS: A Concurrent, Hierarchical and Extensible Fault Simulator," *Proc. Int. Test Conf.*, Philadelphia, PA, pp. 710-716, Nov. 1985.

[20] E. W. Thompson and S. A. Szygenda, "Digital Logic Simulation in a Time-Based, Table-Driven Environment, Part 2, Parallel Fault Simulation," *Computer*, Vol. 8, pp. 38-44, March 1975.

[21] D. B. Armstrong, "A Deductive Method for Simulating Faults in Logic Circuits," *IEEE Trans. Computers*, Vol. C-21, pp. 464-471, May 1972.

[22] E. G. Ulrich and T. Baker, "The Concurrent Simulation of Nearly Identical Digital Networks," *Computer*, pp. 39-44, April 1974.

[23] T. Blank, "A Survey of Hardware Accelerators Used in Computer-Aided Design," *IEEE Design & Test of Computers*, pp. 21-39, August, 1984.

[24] S. J. Hong, "Fault Simulation Strategy for Combinational Logic Networks," *Fault-Tolerant Computing Symp. (FTCS-8) Digest of Papers*, pp. 96-99, June 1978.

[25] M. Abramovici, P. R. Menon, and D. T. Miller, "Critical Path Tracing: An Alternative to Fault Simulation," *IEEE Design & Test of Computers*, Vol. 1, pp. 83-93, February 1984.

[26] F. Brglez and K. Kozminski, "Fast Fault Grading of Sequential Logic," *Proc. Custom Int. Circ. Conf.*, Rochester, NY, pp. 319-324, May 1986.

[27] J. A. Waicukauski, E. B. Eichelberger, D. O. Forlenza, E. Lindbloom, and T. McCarthy, "A Statistical Calculation of Fault Detection Probabilties by Fast Fault Simulation," *Proc. Int. Test Conf.*, Phidelphia, PA, pp. 779-784, Nov. 1985.

[28] P. Goel, "Test Generation Costs Analysis and Projections," *Proc. 17th Design Automation Conference*, Minneapolis, MN, pp. 77-84, June 1980.

[29] V. D. Agrawal, "Sampling Techniques for Determining Fault Coverage in LSI Circuits," *Journal of Digital Systems*, Vol. 5, pp. 189-202, 1981.

[30] D. Harel and B. Krishnamurthy, "Is There Hope for Linear Time Fault Simulation?," *Fault-Tolerant Computing Symp. (FTCS-17) Digest of Papers*, Pittsburgh, PA, pp. 28-33, July 1987.

[31] Y. H. Levendel and P. R. Menon, "Fault Simulation Methods – Extensions and Comparison," *Bell Sys. Tech. Jour.*, Vol. 60, pp. 2235-2258, November 1981.

[32] W. A. Rogers and J. A. Abraham, "A Performance Model for Concurrent Hierarchical Fault Simulation," *Proc. Int. Conf. on CAD*, Santa Clara, CA, pp. 342-345, November 1986.

[33] V. D. Agrawal, S. C. Seth, and P. Agrawal, "Fault Coverage Requirement in Production Testing of LSI Circuits," *IEEE Journal of Solid-State Circuits*, Vol. SC-27, pp. 57-61, Feb. 1982.

[34] S. C. Seth and V. D. Agrawal, "Characterizing the LSI Yield from Wafer Test Data," *IEEE Trans. on CAD*, Vol. CAD-3, pp. 123-126, April 1984, Also in *Proc. Int. Conf. Circ. Comp. (ICCC'82)*, New York, September 1982, pp. 556-559.

[35] T. W. Williams and N. C. Brown, "Defect Level as a Function of Fault Coverage," *IEEE Trans. Computers*, Vol. C-30, pp. 987-988, December 1981.

Chapter 3

TEST GENERATION APPROACHES

Test generation approaches can be classified into three categories: exhaustive, random and algorithmic. For combinational circuits, if the number of primary inputs is small, using exhaustive tests consisting of all possible input vectors to ensure 100% fault coverage is an obvious candidate. Exhaustive techniques can be extended to larger circuits by partitioning into subcircuits such that each subcircuit is tested exhaustively by a reasonably small number of test vectors [1]. However, finding suitable partitions is neither easy nor guaranteed.

Random test generation is a simple and low-cost method in which input vectors are generated randomly. Outputs of the faulty and the fault-free circuits for each random vector are compared by a fault simulator. If any fault is detected, the random vector is retained as a test. The number of random vectors needed for high fault coverage can be very large in some cases. If the number of levels of logic and gate fan-ins are large, this approach becomes less effective. Random test generation techniques can be improved by generating vectors with selected input signal probability [2,3]. Also, the random method can be augmented by deterministic methods [4].

Since early 1960s, numerous algorithms have been proposed for generating test vectors for combinational and sequential circuits. Some of these are effective and widely used in practice; others are of limited practical interest. Most approaches are topological, i.e., they construct the input vector by analyzing the circuit topology. Several famous algorithms for combinational circuits like D-Algorithm [5], PODEM [6] and FAN [7] are widely used and can usually produce satisfactory results. These are commonly known as path sensitization algorithms. Some algebraic methods [8,9,10] have also been reported and are quite elegant and complete. However, their high complexity makes them impractical for large circuits.

Even though several sequential circuit test generators have been developed, their performance is questionable. They have one or more of the following limitations: (1) circuit must be synchronous, (2) limited number of flip-flops, (3) limited number of gates, (4) limited number of vectors per fault, and (5) circuit delays must be neglected.

In the following sections of this chapter, we review the basic concepts of the widely used path sensitization approach. Some representative algorithms for both combinational and sequential circuits are also discussed.

3.1. TEST GENERATION FOR COMBINATIONAL CIRCUITS

An important concept in test generation of combinational circuits is *path sensitization*, which we will illustrate by simple examples.

3.1.1. Path Sensitization Approach

Consider the combinational circuit shown in Fig. 3.1. Test generation for the fault "line c stuck-at-0" is carried out in the following steps:

(1) Set line c to 1. Its value, shown as 1/0, means that it is 1 in good circuit and 0 in faulty circuit.

(2) Justify step 1 by setting line a to 0.

(3) Propagate the value of line c to the output e by setting line d to 1.

In this simple example, the test $0X1$ is found in three steps (X here refers to the don't care state).

Justification, Implication and Propagation. Starting at the site of the fault, path sensitization algorithms try to construct an input vector that will sensitize a path from the fault site to a primary output. In this process, the circuit topology is analyzed gate by gate. A sensitized path is a signal path originating at the fault site such that the signal values along the path uniquely depend on the

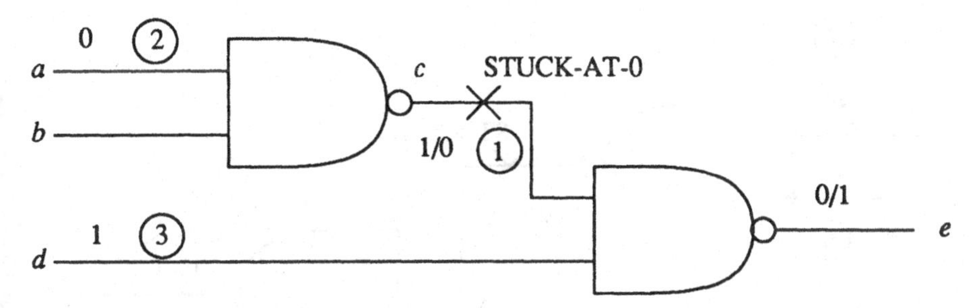

Circled numbers show the sequence of test generation steps.

Fig. 3.1 A simple test generation example.

presence or absence of the fault. In this process, there are three basic operations involved: justification, implication and propagation. Justification means finding a primary input pattern that will produce a specified signal value at an internal node. Step 2 in the above example is justification for setting the node c to 1. When a value is assigned to a signal, it may *imply* other signal values in the circuit. Thus, justification steps are usually interleaved by implications. This will be illustrated in the next example. Propagation is the operation of moving the fault effect forward from the fault site to an output by creating a sensitized path. Step 3 in the above example propagates the fault effect from line c to the output e.

Backtracking. Conflicts can arise during justification and implication operations due to reconvergent fanouts. The backtracking mechanism is used to erase a signal value which causes conflicts and to consider another possible choice. Consider the circuit of Fig. 3.2. Even though this circuit may seem trivially simple, the fault "line c stuck-at-0" is detectable; faulty circuit output function is $e=1$ while the good circuit realizes the function $e = a \cdot b + \bar{a}$.

Test generation steps for this example are:

(1) Set line c to 1. Its state is denoted by 1/0 as before.

(2) Justify $c = 1$ by setting line a to 0.

(3) Carry out forward implication of $a = 0$, i.e., set e to 1.

(4) Propagate the state of c forward. It is impossible. Therefore, backtrack to a previous step where an alternative choice is available.

(5) Backtrack to step 3 and undo it.

(6) Backtrack to step 2, use an alternative choice: set $b = 0$ and leave a unspecified.

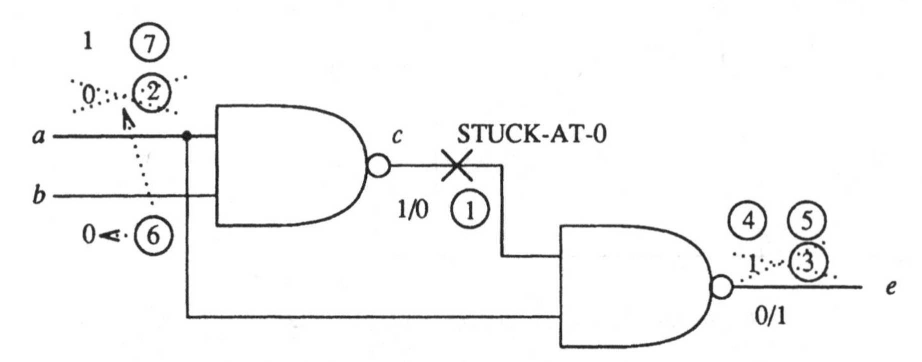

Circled numbers show the sequence of test generation steps.

Fig. 3.2 Test generation with reconvergent fanout.

(7) Propagate state of c forward by setting $a = 1$.

The test, $a = 1$, $b = 0$, is found in seven steps. Although this circuit has about the same size as that in the last example, the test generation took more than twice as many steps. The added complexity is due to the reconvergent fanout of a which made backtracking necessary. In general, large logic circuits contain numerous reconvergent fanouts and the test generation complexity increases rapidly with the size of circuit.

3.1.1.1. D-Algorithm

The D-Algorithm [5] is one of the most popular algorithms used in automatic test pattern generation systems. A five-value $\{0, 1, X$ (unknown), $D, \overline{D}\}$ calculus permits us to carry out the sensitization and justification operations in a formal manner. The symbols D and its complement $\overline{D}$ represent fault effects. If a signal has a value 1 in the fault-free circuit and 0 in the faulty circuit, its value is denoted by D. The complementary situation is denoted by $\overline{D}$. The D-Algorithm consists of three parts, namely, forward implication, D-drive, and backward justification or consistency check. In forward implication, that is similar to logic simulation, output values of logic gates are determined for given input values. In D-drive, the fault effect (D or $\overline{D}$) is propagated toward primary outputs. An exhaustive decision tree is traversed while assigning binary (0 and 1) values to internal nodes of the circuit in a specific order. Consistency checks are performed iteratively traversing another decision tree to justify node values from primary inputs. If any signal value implies a conflict, an alternative value is assigned to the node by performing a backup in the tree traversal. This process is repeated until the effect of the fault reaches a primary output at which point the values on primary inputs correspond to a test. The D-Algorithm propagates fault effects along multiple paths which is essential to guarantee a test.

3.1.1.2. PODEM

The PODEM (Path Oriented DEcision Making) algorithm [6] is generally more effective than D-Algorithm for circuits containing mostly Exclusive-OR gates. Contrary to the D-Algorithm that assigns values to internal signals, PODEM only assigns values to primary inputs. The assignments are determined by first defining an objective of either activating the fault effect or propagating it toward a primary output. If the faulty signal does not have a value D or $\overline{D}$, then the objective would be to generate these values. Once this is accomplished, the next objective is to propagate the D or $\overline{D}$ one level closer to a primary output. This is specified in terms of the values assigned to specific signals. The circuit is traced toward primary inputs to determine the primary input states required to meet the objective. Having assigned a logic value to a primary input, its effect on the circuit is then determined by forward implication. If the objectives are not satisfied, the process repeats from the unsatisfied objective. If conflicts occur, the backtracking only changes primary inputs to untried states.

Conceptually, PODEM solves the test generation problem by applying the branch-and-bound technique in the input vector space. During the test generation process, it creates a binary decision tree as shown in Fig. 3.3. A node here corresponds to a primary input and its two outgoing branches correspond to the 0 and 1 assignments on that primary input. The test generation process finds a path in this binary tree. PODEM conducts a depth-first search. When a new assignment is made to a primary input, a branch is added to the tree. If a conflict occurs due to this assignment, the tree is bounded and the opposite logic value for the last selected primary input is tried, i.e., we trace the brother of the current node. If both values fail, the brother of the ancestor node is examined. Thus, in the worst case, PODEM will exhaustively examine all possible input patterns.

3.1.1.3. FAN Algorithm

Since the backtracking in PODEM can only occur at primary inputs, The total number of backtracks is much reduced over those needed in D-Algorithm. However, there still remain ample possibilities for further reducing the number of backtracks. The FAN algorithm, proposed by Fujiwara and Shimono [7], is a refinement of PODEM that exploits various such possibilities. It performs special processing of fanout points and has been shown to be more efficient and faster than PODEM.

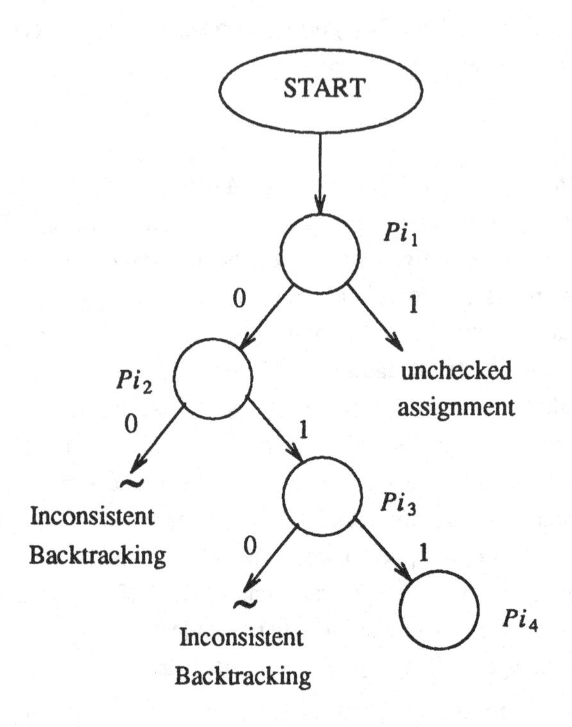

Fig. 3.3 Binary decision tree in PODEM.

3.1.1.4. Subscripted D-Algorithm

The Subscripted D-Algorithm [11] exploits the possibility of concurrently generating test vectors for the output fault and all of the input faults of a gate. Since the D-Algorithm generates test vectors for one fault at a time, it must repeatedly sensitize a path from the output of a gate to a primary output for each fault associated with that gate, even though the paths are almost identical. The main idea in the Subscripted D-Algorithm is to eliminate some of the repeated work.

The subscripted D symbols, D_j, $j = 0, 1, \ldots, n$, are called *flexible signals* and may represent 0 or 1. In order to generate tests for faults associated with an n-input gate, values $D_0, D_1, D_2, \ldots$ and D_n are assigned to the output and inputs, respectively. D_0 is forward-propagated using the normal D-algorithm. After the D_0 has been successfully propagated to a primary output, the D_j, $j = 1, \ldots, n$, are then propagated *backward* to primary inputs. In this back-propagation process, the signal values are represented and manipulated symbolically using D_j's to concurrently handle several independent paths. Once the symbols are successfully propagated to primary inputs, a set of test vectors for the output fault and all of the gate input faults is determined by inspecting the symbols at primary inputs.

It can be shown that the computation effort of generating the *symbolic* test vector in the subscripted D-algorithm is about the same as that of generating a test vector for a single fault in D-algorithm. Therefore, the effectiveness of subscripted D-algorithm is enhanced by the amount of fan-ins present in the circuit.

3.1.1.5. CONT Algorithm

The objective of CONT (A CONcurrent Test generation algorithm) [12] is also to concurrently generate tests for a set of faults. For a given fault, CONT follows the PODEM algorithm. In addition, however, during forward implication, lists of active faults are generated for every gate. The fault list of a gate contains the faults whose effects appear at the output of that gate in the presence of the current primary input values. Thus, the forward implication operation is similar to concurrent fault simulation discussed in the last chapter. Once a test vector is found for the given fault, all faults in the fault lists associated with primary outputs are also detected by this vector. Also, when a conflict occurs, instead of changing the primary input values, the target is switched to another fault that is activated by the current vector and is most likely to be propagated to a primary output. In this case, the fault lists of gates are useful in selecting the new target fault. The strategy of switching target fault reduces the total number of backtrackings.

Efficient test generation programs for combinational circuits often incorporate a dynamic selection between several heuristics [13, 14].

3.1.2. Boolean Difference Approach

The Boolean difference approach [8] is an algebraic technique of test generation. Even though this approach does not have the popularity enjoyed by the path sensitization approaches, it is elegent and provides a good insight into the process of test generation.

Let $F(X)$ be the function of the fault-free circuit and $F'(X)$ the function of the circuit in the presence of a given fault. The set $\{ X \mid F(X) \oplus F'(X) = 1 \}$, as defined below, is the set of all input vectors that distinguish between the two functions. This is the complete set of test vectors for the given fault. The *Boolean difference* of logic function $F(x_1, x_2, \ldots, x_n)$ with respect to an input variable x_i is defined as:

$$\frac{dF(X)}{dx_i} = F(x_1, \ldots, x_i, \ldots, x_n) \oplus F(x_1, \ldots, \overline{x}_i, \ldots, x_n)$$

Since x_i can assume values 1 or 0, this Boolean difference can be expressed as:

$$\frac{dF(X)}{dx_i} = F(x_1, \ldots, x_{i-1}, 1, x_{i+1}, \ldots, x_n)$$

$$\oplus F(x_1, \ldots, x_{i-1}, 0, x_{i+1}, \ldots, x_n)$$

$$= F_i(1) \oplus F_i(0)$$

Since $F(X)$ is known, $F_i(1)$ and $F_i(0)$ can be easily derived.

Consider an input fault, x_i stuck-at-0. The faulty function is $F_i(0)$. Therefore, all tests for the fault can be determined by solving

$$F(X) \oplus F_i(0) = 1.$$

Using Boolean algebra, we can express the above equation by the input variable x_i and the Boolean difference with respect to x_i as:

$$x_i \frac{dF(X)}{dx_i} = 1.$$

Similarly, the complete set of tests for x_i stuck-at-1 can be expressed by

$$\{X \mid \overline{x}_i \frac{dF(X)}{dx_i} = 1\}.$$

The complete set of tests for stuck-at faults on any internal signal I in the circuit can also be generated similarly if F is expressed in terms of X and I. Furthermore, by defining multiple Boolean differences, tests for multiple stuck-at faults can also be derived [15]. However, the manipulation of algebraic equations is complex and makes this class of approaches impractical for large circuits.

3.2. TEST GENERATION FOR SEQUENTIAL CIRCUITS

The response of a sequential circuit depends on its primary inputs and the stored internal states. The stored states can retain their value over time. Thus, combinational test generation methods can be applied to sequential circuits if the element of *time* is introduced.

3.2.1. Iterative Array Approach

Many sequential circuit test generators have been devised on the basis of the fundamental combinational algorithms. A combinational model for a sequential circuit is constructed by regenerating the feedback signals from *previous-time* copies of the circuit. Thus the timing behavior of the circuit is approximated by combinational levels. Topological analysis algorithms that activate faults and sensitize paths through these multiple copies of the combinational circuit are used to generate tests.

We illustrate this approach by a simple example. Consider the latch in Fig. 3.4. Let us assume that all gates have zero delay. The result of applying the combinational test generation approach is shown in the figure. The process stops with inputs 10 and the output 1. Since the output is the same in both good and faulty circuits, a test is not found.

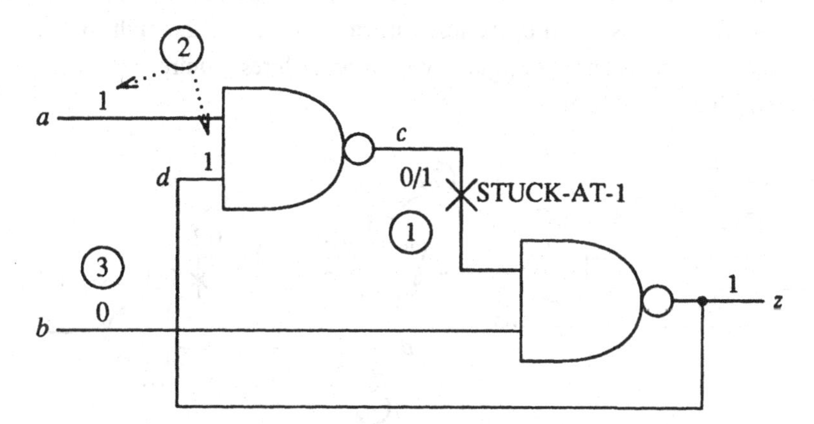

Circled numbers show the sequence of test generation steps.

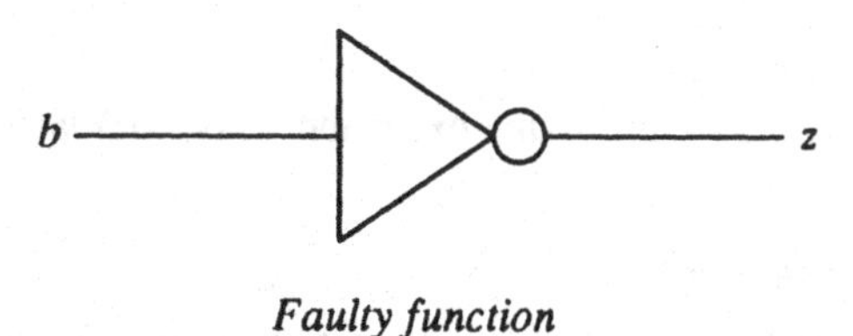

Faulty function

Fig. 3.4 A NAND latch example.

Figure 3.4 also shows that the faulty function, when "line c is stuck-at-1", is $z = \overline{b}$. The fault blocks the feedback and the circuit no longer has the storing capability. It can be easily detected by first applying a 10 input and then following it up by 11. The first pattern will produce a 1 output irrespective of the fault. The second pattern simply stores the state of the latch. In the good circuit, the output will remain as 1 while it will change to 0 in the faulty circuit.

Our combinational test generator was not able to solve this problem due to the zero-delay assumption. If we consider finite delays of gates, it is possible to first apply the value of $z=1$ to d and then change b to 1 to sensitize the path for the fault. A common practice is to cut the feedback path. This is shown in Fig. 3.5. A copy of the circuit is attached to generate the feedback signal d. The test generation begins from the copy shown as *current time frame*. In general, any number of time frames (previous or future) can be added on either side of the current time frame. However, the complexity of the model increases with the number of time frames. Another problem occurs due to the signals left unspecified by the test generator. For example, the test in Fig. 3.5 is a 11 pattern preceded by $X0$ where X denotes the don't care state. If we set X to 1, we get the desired test. But if X is set to 0, the test will cause a *race* in the fault-free circuit. Thus, sequential circuit tests generated by such procedures require special processing to avoid timing problems.

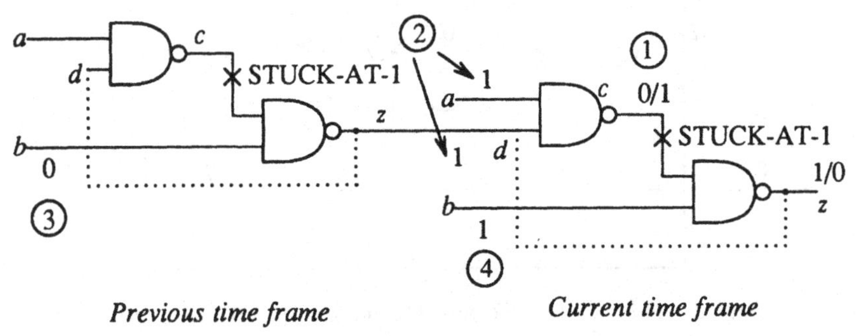

Fig. 3.5 Time frame extension of NAND latch.

3.2.1.1. Extended D-Algorithm

A general model for synchronous sequential circuits is shown in Fig. 3.6. The output is a logic function of primary inputs and the present state of flip-flops. This circuit can be modeled by a combinational network as shown in Fig. 3.7 [16, 17]. Feedback lines are cut and the combinational portion is duplicated. The block CC_n is the nth copy of the combinational portion and corresponds to the time frame n. The inputs of CC_n include the primary inputs I_n and the *pseudo inputs* SI_n obtained from the feedback signals produced by CC_{n-1}. Similarly, O_n and SO_n are the primary outputs and pseudo outputs of the time frame n.

In order to find a test sequence to detect a single stuck-at fault, we must first determine the number of time frames needed in the combinational model. The D-Algorithm is then applied to the jth copy of the combinational model, called the

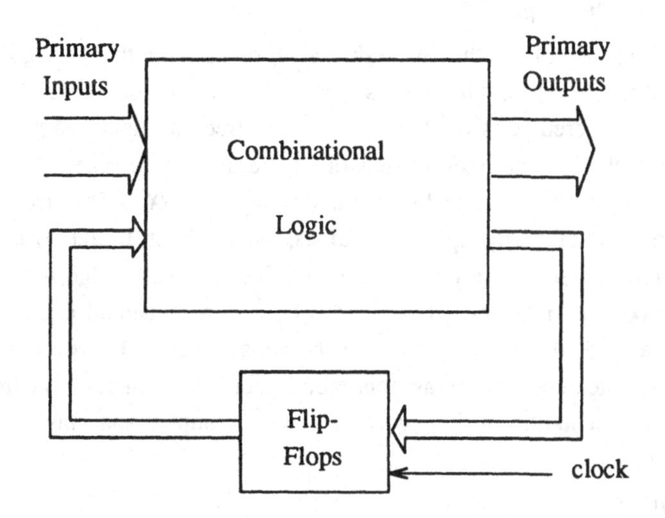

Fig. 3.6 Model of synchronous sequential circuits.

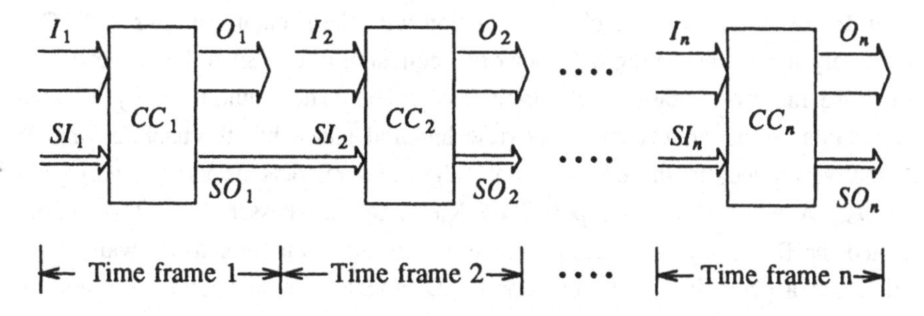

Fig. 3.7 Iterative combinational model.

current time frame. If the D-Algorithm assigns a value to any pseudo input, this assignment is then justified in the $(j-1)$th copy. Similarly, if the fault effect is propagated to a pseudo output, propagation must continue into the $(j+1)$th and subsequent copies until the fault effect reaches a primary output. It is worth noting that a single stuck-at fault in the sequential circuit corresponds to a multiple fault in the combinational model, since the complete model contains one fault per time frame. Therefore, the D-Algorithm applied to the combinational model should be modified to deal with fault multiplicity. The multiple fault consideration and the extra complexity of the replicated logic make this approach unrealistic for large circuits.

3.2.1.2. Nine-Value Algorithm

Muth [18] extended the five-value algebra used in the D-Algorithm to a nine-value algebra. These nine values (1/1, 1/0, 1/X, 0/1, 0/0, 0/X, X/1, X/0, and X/X) represent ordered pairs of states of the fault-free and faulty circuits. The five values used in the D-Algorithm are actually a subset of these nine values; a 1 used in D-Algorithm is equivalent to 1/1, a 0 is 0/0, an X is X/X, a D is 1/0, and a $\bar{D}$ is 0/1. The additional partly specified values, 0/X, 1/X, X/1 and X/0, provide a greater degree of freedom in test generation. The nine-value algorithm takes into account the possible repeated effects of the fault in a sequential circuit and a test can be guaranteed, if one exists, for synchronous circuits. However, the method does not guarantee success for asynchronous circuits. Even for synchronous circuits, the implementation of the algorithm is very complex and may be inefficient for large circuits.

3.2.1.3. SOFTG

Primarily, SOFTG [19] is a sequential version of PODEM. Tests are constructed by tracing backward through the circuit in much the same way as a combinational test generator, but all forward signal propagation is carried out by an event-driven simulator. The close interaction with the simulator allows SOFTG to effectively model the timing behavior of a sequential circuit such that the tests will not cause races or hazards in the fault-free circuit. The method is very effective for circuits of moderate complexity (few hundred gates) but the backtracking for alternative choices in the combinational algorithm restricts its use for very large circuits. A similar idea is reported by Kjelkerud & Thessen [20]. They implemented the D-Algorithm by using a table-driven logic simulator for forward implication and a deductive fault simulator for D-propagation. The fault simulator can ensure that the vectors are hazard-free.

3.2.1.4 Backtrace Algorithms

All sequential test generation algorithms discussed above have a process flow that is bidirectional in time where time refers to the *time* of events in the circuit. This is because the starting event is the fault activation at the site of the fault. In general, the detection will take place at a future time while the primary inputs must assume their values in the past. Marlett [21] proposed a unidirectional single path sensitization algorithm which works backward both spatially and in time from a selected output toward the fault. For a given fault, a path and an output pin with best observability are first selected. The test generator begins by setting up sensitizing conditions at the path output and proceeds backward. The unidirectional (backward) time flow simplifies implementation. However, it might lose some accuracy because only one path is allowed to be sensitized at a time.

This algorithm is generally known as the extended backtrace (EBT) and its implementations [22,23] have produced practical test generation programs.

While the EBT algorithm sensitizes a preselected path, the recently proposed BACK algorithm [24] only preselects a primary output to which it sensitizes all paths starting at the fault site.

3.2.2. Verification-Based Approaches

The verification approach relies on determining whether or not the circuit under test is operating in accordance with its state table. This requires finding a sequence which forces the circuit to go through all states and output transitions [25,26]. Actually, this approach can be considered a form of functional testing. Unfortunately, the high complexity of building the complete state table limits its application for large circuits. In addition, test sequences are usually excessively long for highly sequential circuits. However, the technique has been successfully applied to protocol testing [27].

3.2.2.1. SCIRTSS

SCIRTSS (Sequential CIRcuit Test Search System) [28] divides the test generation process into three steps. The D-Algorithm is first applied to the combinational portion of the circuit. The test vector, thus obtained, is then split between primary inputs and the *present* state. At this point, the problem becomes one of finding an input sequence to bring the circuit to this state and finding another sequence to propagate the stored fault effect to a primary output. These two steps require finding suitable paths through the state graph. In order to limit the effort, the searches for these paths are only conducted over a restricted state diagram and a small portion of the much larger set of data states. The data states included in search depend on user-supplied heuristic information.

A test generator for finite state machines reported by Ma et al [29] applies PODEM with iterative-array type of processing for forward propagation. The backward justification phase is replaced by a procedure that attempts to search for a path in the state transition graph from a given reset state to the present state required by PODEM. In order to limit the complexity of the state transition digram, a *partial* state diagram is constructed containing all valid states of the finite state machine but only a few transition edges. In searching for a path through the state transition graph (STG), the STGs of faulty circuits are assumed to be the same as that of the fault-free circuit. Since this assumption may not be universally true, the generated vectors must be verified by a fault simulator.

3.2.3. Functional and Expert System Approaches

A functional approach [30, 31], uses a high-level description of the circuit, to generate test sequences that will verify whether the designed functions are being performed correctly. This approach often uses restricted fault models like the line stuck faults on inputs and outputs of functional blocks or higher-level fault models such as errors in the truth table of a combinational block, or a change in the state table of a sequential functional block. The functional testing is often the method of choice for testing very large circuits like microprocessors. However, it is difficult to evaluate the quality of functional test vectors. In many cases, these tests may not be capable of detecting every possible failure that can occur.

The expert-system approachs [32, 33] incorporate the knowledge of human test programmers and a variety of algorithmic and heuristic techniques into an interactive software environment. These approachs require intensive user interaction, higher-level modeling libraries, and often restricted architectures. For these reasons, it has been difficult to effectively integrate them into existing design methodologies.

REFERENCES

[1] E. J. McCluskey and S. Bozorgui-Nesbat, "Design for Autonomous Test," *IEEE Trans. Comp.*, Vol. C-30, pp. 860-875, Nov. 1981.

[2] P. Agrawal and V. D. Agrawal, "On Monte Carlo Testing of Logic Tree Networks," *IEEE Trans. Comp.* , Vol. C-25, pp. 664-667, June, 1976.

[3] R. Lisanke, F. Brglez, A. J. DeGeus, and D. Gregory, "Testability-Driven Random Test-Pattern Generation," *IEEE Trans. on CAD*, Vol. CAD-6, pp. 1082-1087, Nov. 1987.

[4] V. D. Agrawal and P. Agrawal, "An Automatic Test Generation System for Illiac IV Logic Boards," *IEEE Trans. Comp.*, Vol. C-21, pp. 1015-1017, September 1972.

[5] J. P. Roth, W. G. Bouricius, and P. R. Schneider, "Programmed Algorithms to Compute Tests and to Detect and Distinguish Between Failures in Logic Circuits," *IEEE Trans. on Electronic Computers*, Vol. EC-16, pp. 567-580, October 1967.

[6] P. Goel, "An Implicit Enumeration Algorithm to Generate Tests for Combinational Logic Circuits," *IEEE Trans. Comp.*, Vol. C-30, pp. 215-222, March 1981.

[7] H. Fujiwara and T. Shimono, "On the Acceleration of Test Generation Algorithms," *IEEE Trans. Comp.*, Vol. C-32, pp. 1137-1144, Dec. 1983.

[8] F. F. Sellers, M. Y. Hsiao, and L. W. Bearnson, "Analyzing Errors with Boolean Difference," *IEEE Trans. Comp.*, Vol. C-17, pp. 676-683, July 1968.

[9] D. B. Armstrong, "On Finding a Nearly Minimal Set of Fault Detection Tests for Combinational Logic Nets," *IEEE Trans. Electronic Computers*, pp. 66-74, October 1965.

[10] K. Kinoshita, Y. Takamatsu, and M. Shibata, "Test Generation for Combinational Circuits by Structure Description Functions," *Fault-Tolerant Computing Symp. (FTCS-10) Digest of Papers*, pp. 152-154, October 1980.

[11] J. F. McDonald and C. Benmehrez, "Test Set Reduction Using the Subscripted D-Algorithm," *Proc. 1983 Int. Test Conf.*, pp. 115-121, Oct. 1983.

[12] Y. Takamatsu and K. Kinoshita, "CONT: A Concurrent Test Generation Algorithm," *Fault-Tolerant Computing Symp. (FTCS-17) Digest of Papers*, Pittsburgh, PA, pp. 22-27, July 1987.

[13] M. Abramovici, J. J. Kulikowski, P. R. Menon, and D. T. Miller, "SMART and FAST: Test Generation for VLSI Scan Design Circuits," *IEEE Design & Test of Computers*, Vol. 3, pp. 43-54, August 1986.

[14] M. H. Schulz and E. Auth, "Advanced Automatic Test Pattern Generation and Redundancy Identification Techniques," *18th Fault-Tolerant Computing Symp. (FTCS-18) Digest of Papers*, Tokyo, Japan, pp. 30-35, June 1988.

[15] C. T. Ku and G. M. Masson, "The Boolean Difference and Multiple Fault Analysis," *IEEE Trans. Comp.*, Vol. C-24, pp. 62-71, Jan. 1975.

[16] H. Kubo, "A Procedure for Generating Test Sequences to Detect Sequential Circuit Failures," *NEC J. Res. Dev.*(12), pp. 69-78, October 1968.

[17] G. R. Putzolu and J. P. Roth, "A Heuristic Algorithm for the Testing of Asynchronous Circuits," *IEEE Trans. Comp.*, Vol. C-20, pp. 639-647, June 1971.

[18] P. Muth, "A Nine-Valued Circuit Model for Test Generation," *IEEE Trans. Comp.*, Vol. C-25, pp. 630-636, June 1976.

[19] T. J. Snethen, "Simulator Oriented Fault Test Generator," *Proc. 14th Des. Auto. Conf.*, New Orleans, Louisiana, pp. 88-93, June 1977.

[20] E. Kjelkerud and O. Thessen, "Generation of Hazard Free Tests using the D-Algorithm in a Timing Accurate System for Logic and Deductive Fault Simulation," *Proc. Des. Auto. Conf.*, San Diego, CA, pp. 180-184, June

1979.

[21] R. A. Marlett, "EBT: A Comprehensive Test Generation Technique for Highly Sequential Circuits," *Proc. Des. Auto. Conf.*, Las Vegas, Nevada, pp. 335-339, June 1978.

[22] S. Mallela and S. Wu, "A Sequential Circuit Test Generation System," *Proc. Int. Test Conf.*, Philadelphia, PA, pp. 57-61, November 1985.

[23] R. A. Marlett, "An Effective Test Generation System for Sequential Circuits," *Proc. Des. Auto. Conf.*, Las Vegas, Nevada, pp. 250-256, June 1986.

[24] W. T. Cheng, "The BACK Algorithm for Sequential Test Generation," *Proc. Int. Conf. Computer Design (ICCD-88)*, Rye Brook, NY, pp. 66-69, October 1988.

[25] F. C. Hennie, "Fault Detecting Experiments for Sequential Circuits," *Proc. 5th Annual Symp. on Switching Circuit Theory and Logic Design, New Jersey*, pp. 8-22, Nov. 1964.

[26] P. G. Kovijanic, "A New Look at Test Generation and Verification," *Proc. 14th Des. Auto. Conf.*, New Orleans, Louisiana, June 1977.

[27] A. V. Aho, A. T. Dahbura, D. Lee, and M. U. Uyar, "An Optimization Technique for Protocol Conformance Test Generation Based on UIO Sequences and Rural Chinese Postman Tours," *Proc. 8th Int. Symp. on Protocol Specification, Testing, and Verification*, Atlantic City, NJ, June 1988.

[28] F. J. Hill and B. Huey, "A Design Language Based Approach to Test Sequence Generation," *Computer*, Vol. 10, pp. 28-33, June 1977.

[29] H.-K. T. Ma, S. Devadas, A. R. Newton, and A. Sangiovanni-Vincentelli, "Test Generation for Sequential Finite State Machines," *Int. Conf. Computer-Aided Design (ICCAD '87)*, pp. 288-291, Nov. 1987.

[30] T. Sridhar and J. P. Hayes, "A Functional Approach to Testing Bit-Sliced Microprocessors," *IEEE Trans. Comp.*, Vol. C-30, pp. 563-571, Aug. 1981.

[31] D. Brahme and J. A. Abraham, "Functional Testing of Microprocessors," *IEEE Trans. Comp.*, Vol. C-33, pp. 475-485, June 1984.

[32] M. J. Bending, "Hitest: A Knowledge-Based Test Generation System," *IEEE Design & Test of Computers*, Vol. 1, pp. 83-92, May 1984.

[33] N. Singh, *An Artificial Intelligence Approach to Test Generation*, Kluwer Academic Publishers, Norwell, MA, 1987.

Chapter 4

SIMULATION-BASED DIRECTED-SEARCH APPROACH

It is a normal practice to verify tests through simulation. This is done with the help of a fault simulator that checks the timing behavior (races, hazards, etc.), computes the circuit response, and verifies detectability of modeled faults. In the last chapter, we mentioned several test generators that use simulators in the forward implication or fault propagation phase [1,2,3]. However, backtracing through the circuit structure was still an essential part of the process. In this context, the role of reconvergent fanouts in the circuit is well known. For example, a signal value set to propagate the fault effect through a gate may fanout to other gates and block further propagation of the same fault effect. To overcome such difficulties, test generators must *backtrack* and make alternate signal choices. Our objective in developing a new method, therefore, is to completely eliminate any backtrace and rely on simulation for all decisions.

4.1. PRINCIPLE OF DIRECTED SEARCH

The simplest simulation-based test generator is perhaps the random vector method. However, the efficiency of this method, especially for sequential circuits, is generally low. We will, therefore, introduce the concept of directed-search for test generation. The basic concept of directed-search is as follows: Suppose we wish to generate a test for a given fault. Also suppose we have an input vector; a previous test or a random vector will suffice. We perform fault simulation. From the simulation result, a cost function, that, in some way, determines how *far* the vector is from being a test, is computed. The cost should be so defined that it will be below a threshold only if the simulated vector detects the target fault. If the fault is not detected, i.e., the cost is high, then its reduction through suitable changes in the vector leads to a test. This approach is a combination of (1) simulation, and (2) cost function guidance.

There are several advantages of using simulation: (1) it involves no backtracking, (2) the event-driven simulation deals with circuit delays in a very natural way, and (3) even asynchronous sequential circuits can be handled.

The Role of Cost function. A directed search can be conducted in the input vector space by defining an appropriate cost function. The cost function $C_f(V_i)$ is computed for an input vector V_i and for a given fault f, and depends only on the results of simulation of the good and the faulty circuits. The lower the cost, the closer the input vector is to being a test. A properly defined cost will be lower than some cost threshold C_0 if, and only if, the input vector detects the fault.

Starting with any vector, the circuit is simulated and the cost is computed for all vectors that are at unit Hamming distance from this vector. We select the vector with minimum cost. Successive moves either lead to a test or the search terminates at a local cost minimum. In the latter case, the search process can be restarted with a new initial vector.

It is not easy to sketch the behavior of cost function in the multi-dimensional input vector space. If we simply use a single dimension to represent the input vector space, the curve of cost function may be as shown in Figs. 4.1 and 4.2, where C_i is the cost of the initial vector and C_0 is the cost threshold. The search will be successful if the curve is smooth like the one in Fig. 4.1. A local cost minimum will be reached for the case in Fig. 4.2. However, a test can still be found if the search is started from another inital vector as shown.

Since the cost function is entirely derived from the results of simulation, circuit delays and timing problems (races, hazards, etc.) are taken into account by the simulator. Also, all decisions during the search are made on the basis of inputs

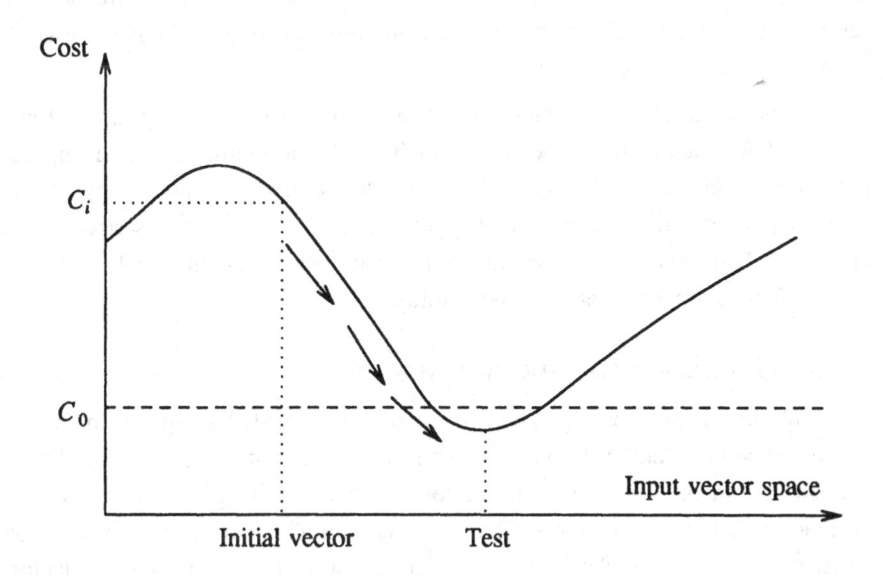

Fig. 4.1 Curve of cost function.

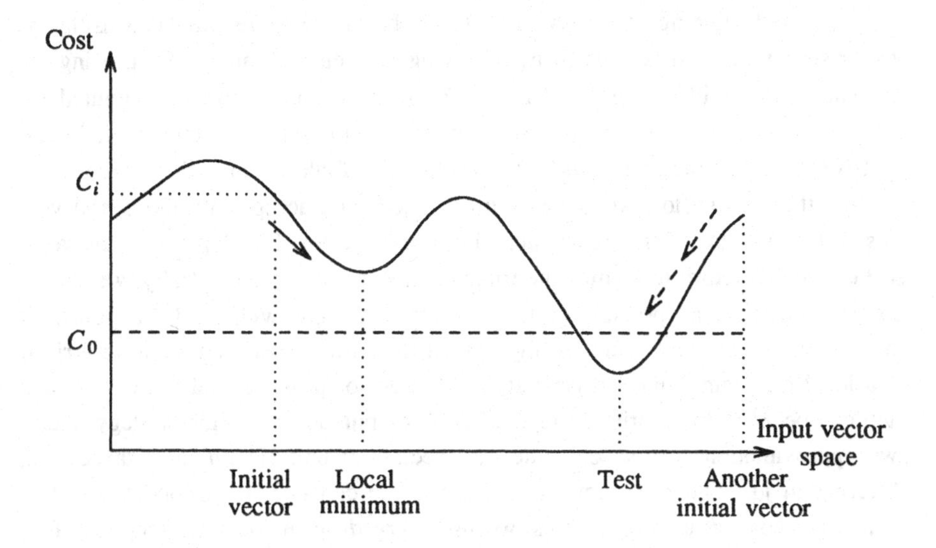

Fig. 4.2 Local cost minimum.

and outputs and no explicit consideration of the internal structure (reconvergent fanouts) is required. Furthermore, asynchronous sequential circuits present no problem for the event-driven simulation.

It is the guidance that makes the search process significantly more efficient compared to random test generation which is also a simulation-based approach. One way to derive this guidance is to use testability that may be dynamically evaluated for specific vectors. Such guidance methods will be described in later chapters. Test generation programs, employing static testability analysis for guidance often find the advantage to be marginal [4].

4.2. AN OVERVIEW OF THE NEW METHOD

We use a multi-pass process. Each pass is executed as shown in Fig. 4.3. We begin with a fault list and a vector set. In the first pass, the fault list may contain all stuck faults and the vector set may be empty. Alternatively, at designer's option, the vector set may contain initialization vectors or functional verification vectors. If the vector set is empty, it is then initialized by placing a randomly generated or some arbitrary vector in it. In subsequent passes, the fault list will contain the faults that were classified as *hard to detect* during the previous pass. Also, the vector set will be the vector set at the end of the preceding pass.

At the beginning of a pass all faults in the fault list are simulated using the vector set and the list is updated by removing the detected faults. Next, using the procedures that will be described in the following chapters, cost is computed for the undetected faults at the *last vector* in the vector set. The fault with lowest cost is selected as the *target fault*. The cost of this fault is called the *current cost*. To find the next vector, we generate *trial vectors* and accept only those trial vectors that reduce the cost. Before applying a trial vector, flip-flop states are saved so that we can return to them if the trial vector is not accepted. Ideally, we should compute costs for all possible vectors and select the one with the lowest cost. If the cost of the selected vector is higher than the current cost then we have arrived at a local cost minimum. In general, the number of possible trial vectors at each step can be 2^n-1 for n primary inputs. Our computation reduction strategy makes two approximations. First, only the trial vectors at unit Hamming distance (i.e., differing in just one bit) from the last vector are considered. Second, instead of computing costs at all trial vectors, we use a *greedy* heuristic and accept the first trial vector having a cost lower than the last vector. If all such trial vectors have higher costs then the target fault is removed from the fault list and placed in the list of *hard to detect* faults to be considered in subsequent passes. A pass can

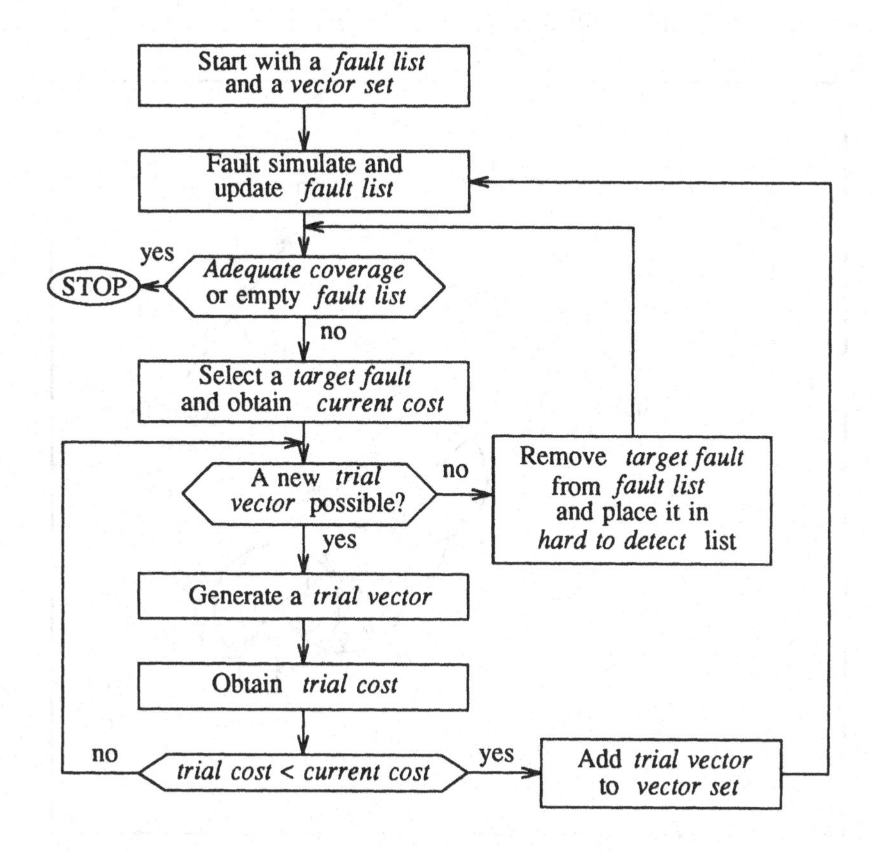

Fig. 4.3 A flow chart of the directed search method.

terminate at one of two conditions: (1) adequate fault coverage is achieved, or (2) every fault initially in the fault list is either detected or placed in the hard to detect list, i.e., the fault list becomes empty.

Notice that in the flow chart of Fig. 4.3, when a trial vector is added to the vector set, the process of selecting a target fault (the fault with lowest cost) is repeated. If the new vector has only lowered the cost of the previous target fault but has not detected it then it is possible to select a different target fault. This *target switching* is further discussed in Chapter 6.

Fig. 4.3 gives an overview of the new method. The algorithms are slightly different for handling combinational circuits, synchronous sequential circuits and asynchronous sequential circuits. For example, for combinational circuits, the final vector at which the cost drops below C_0 is the test. For sequential circuits, all vectors from the initial to the final vector form the test sequence. The details of various algorithms will be discussed in Chapters 6 and 7.

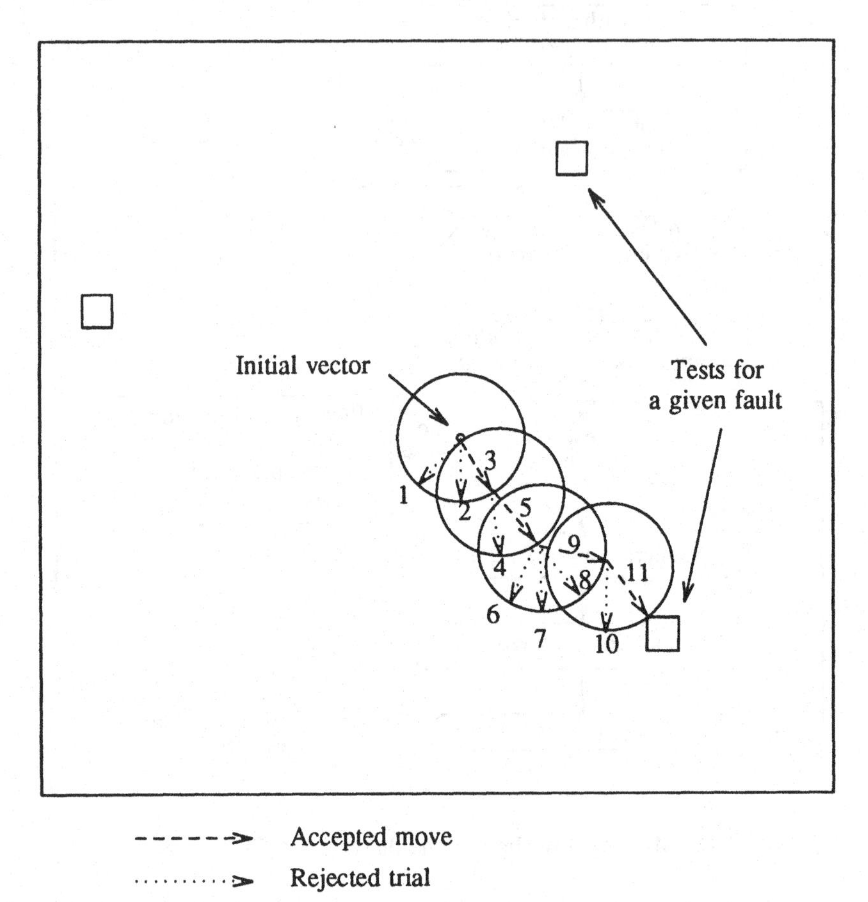

Fig. 4.4 A directed search in input vector space.

Fig. 4.4 illustrates a test search in the input vector space. The three points shown by squares are the possible tests for the given target fault; the large circles represent the unit Hamming distance neighborhood of accepted vectors. The dotted arrows show the trial vectors that are simulated but not accepted because of their higher costs. The dashed arrows indicate the accepted moves. The numbers associated with arrows show the sequence of trials in the test generation process. In this illustration, the first two trial vectors, 1 and 2, have costs higher than the initial vector and hence they are rejected. The third vector reduced the cost and is accepted. The search continues and after simulating eleven vectors, a test is found.

4.2.1. The Unit Hamming Distance Heuristic

In a directed search, before deciding on a move, one normally considers the cost at all neighboring points in the search space. We have defined the neighborhood of a given vector as all vectors at unit Hamming distance (i.e., differing in just one bit) from it. This seemingly arbitrary choice is similar to computing *partial differences* with respect to the input variables for a steepest descent solution. In the present situation, however, a test can be guaranteed if the neighborhood includes vectors that differ in 1, 2, . . . , n bits where n is the number of primary inputs. Thus, our heuristic corresponds to considering only n of these 2^n neighbors. The judgement on whether we have achieved a practical trade-off in reducing computational complexity without giving up accuracy should be based on the experimental results that will be presented in Chapters 6 and 7.

4.2.2. Cost Functions

The cost function is computed for an input vector and a specific fault. We consider two different methods of deriving cost functions. In the first method, the cost is defined as follows:

$$C(V) = \min_{i \in PO} \frac{1}{|\ F_i(V) - F'_i(V)\ |}$$

where $F_i(V)$ is the output value at the ith primary output for the input vector V, $F'_i(V)$ is the corresponding output value in the faulty circuit, and PO is the set of primary outputs. When the two output values differ, the cost will be 1.0 and the fault is detected. If the output values are identical, i.e., the fault is not detected, then cost will be infinity. If we have two vectors, V_i and V_j, such that neither of them detects the fault under consideration, then the cost for each vector will be infinity. Thus the cost as defined above will not allow us to discriminate between these vectors. However, to conduct a directed search for a test we would like to know which vector is better. In the next chapter, we will describe a new simulation technique, called the *threshold−value simulation*, that can accurately perform logic simulation and also discriminate between logically identical signal values to provide useful data for computation of a cost function.

In an alternative method, the cost functions are not only computed from the signal values at the primary outputs but the internal signal states are also used. One advantage of this method is that it can use a concurrent fault simulator for test generation. The cost is a function of the *distance* of the fault effect from primary outputs. The details will be discussed in Chapter 7.

REFERENCES

[1] T. J. Snethen, "Simulator Oriented Fault Test Generator," *Proc. 14th Des. Auto. Conf.*, New Orleans, Louisiana, pp. 88-93, June 1977.

[2] E. Kjelkerud and O. Thessen, "Generation of Hazard Free Tests using the D-Algorithm in a Timing Accurate System for Logic and Deductive Fault Simulation," *Proc. Des. Auto. Conf.*, San Diego, CA, pp. 180-184, June 1979.

[3] Y. Takamatsu and K. Kinoshita, "CONT: A Concurrent Test Generation Algorithm," *Fault-Tolerant Computing Symp. (FTCS-17) Digest of Papers*, Pittsburgh, PA, pp. 22-27, July 1987.

[4] A. Ivanov and V. K. Agarwal, "Testability Measures - What Do They Do for ATPG?," *Proc. IEEE International Test Conf.*, pp. 129-138, September 1986.

Chapter 5

THRESHOLD-VALUE SIMULATION

In this chapter, we present a new method of logic simulation with implicit evaluation of signal controllabilities. To make such simulation possible we have developed a model for logic gate known as the *threshold-value model.*

Consider the two cases shown in Fig. 5.1. The logical output of the AND gate is 0 in both cases (the values in parentheses will be explained later.) The logic value of signal D in (a) can be changed to 1 by changing just one input (i.e., C) to 1. In (b), on the other hand, all inputs (A, B, and C) must be changed to make the output 1. Qualitatively, we express this difference between the two cases by saying that the 0 at the output in (b) is *stronger*. The strength here is related to the effort needed to change the value. This information, as we will show, can be used for computing the cost function to direct the search for a test. Since the conventional logic simulation carries no information about such strengths we need a new method of simulation [1,2] in our application.

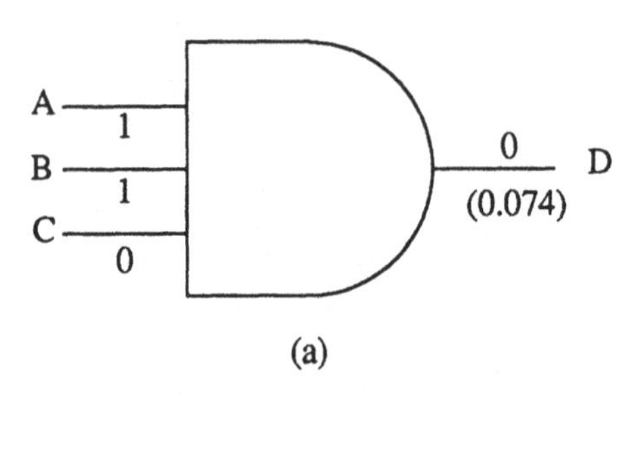

(a)

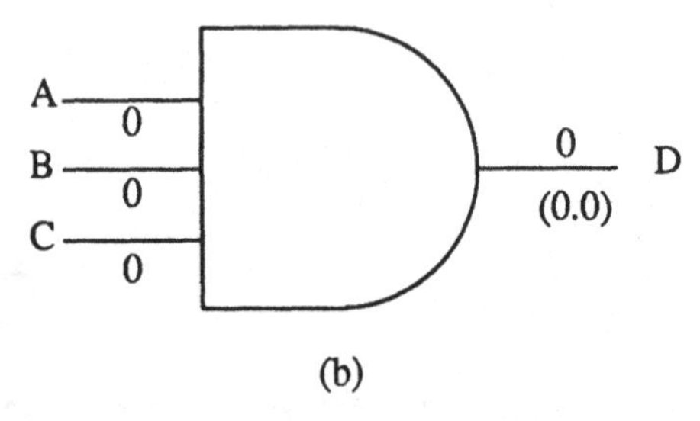

(b)

Fig. 5.1 Examples for threshold-value simulation.

5.1. THRESHOLD-VALUE MODEL

Suppose we denote the value of line i as V_i and compute an average input value for an n-input gate as

$$V = \frac{1}{n} \sum_{i=1}^{n} V_i$$

In Fig. 5.1(a), $V_A = 1$, $V_B = 1$, and $V_C = 0$. Therefore, $V = 0.667$. Next, we define the function of an AND gate by $T_{AND}(V)$ as shown in Fig. 5.2. For the purpose of logic simulation, we interpret the values between 0.0 and 0.1 as logic 0 and those between 0.9 and 1.0 as logic 1. In terms of dynamic controllability, the value 0.0 is interpreted as the *strongest* 0 and 0.1 as the *weakest* 0. The outputs shown in parentheses in Figs. 5.1 (a) and (b), 0.074 and 0.0, respectively, will both be interpreted as logic 0. However, the fact that changing the value of signal D to 1 is easier in (a) than in (b) is indicated by the slightly higher output value in (a).

We call T_{AND} the threshold function of AND gate. The threshold function, T_{OR}, for an OR gate is shown in Fig. 5.3. The threshold values, 0.1 and 0.9, are chosen according to the maximum gate fanin in the circuit as described below.

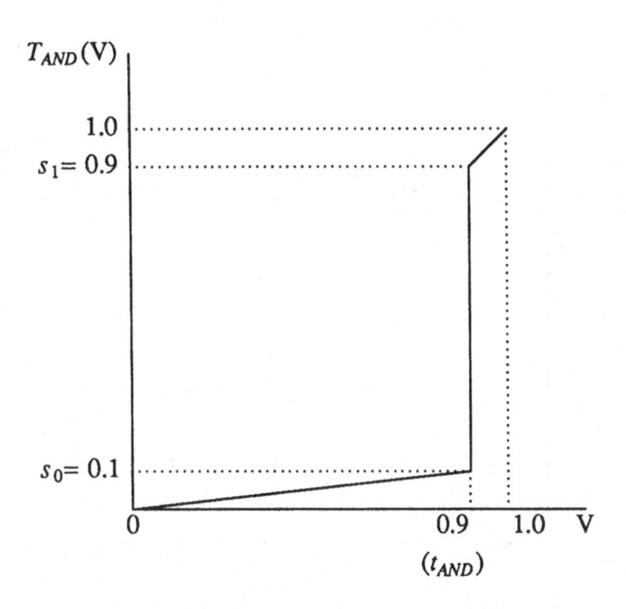

Fig. 5.2 Threshold function for AND gate (T_{AND}).

5.2. DETERMINATION OF THRESHOLDS

We have used two *signal thresholds,* s_0 and s_1, as shown in Figs. 5.2 and 5.3. Signal values in the range $[0,s_0]$ are equivalent to logic 0 and those in the range $[s_1,1]$ are equivalent to logic 1. We define the *gate threshold* t_G for gate G as the average input value at which the gate output switches. Using the conditions that an AND gate must not switch to 1 as long as at least one input is at logic 0 and that it must switch to 1 when all inputs are at logic 1, we obtain the following relation:

$$1 - \frac{1-s_0}{n_{\max}} < t_{AND} \leq s_1$$

where $n_{\max}$ is the maximum fanin in the circuit. More specifically, s_1 is the minimum average input value with output at logic 1 and $1 - \frac{1-s_0}{n_{\max}}$ is the maximum average input value with output at logic 0. Fig. 5.4 illustrates these two extreme cases bounding the threshold, t_{AND}. If the width of the logic 0 range is made equal to that of the logic 1 range, then

$$s_0 = 1 - s_1.$$

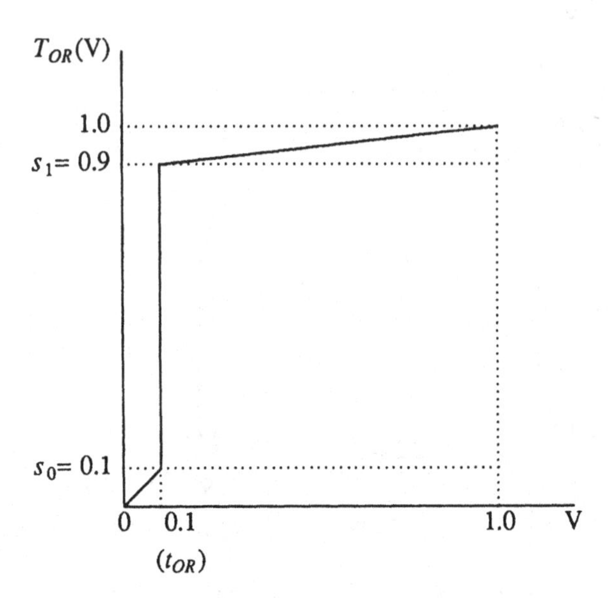

Fig. 5.3 Threshold function for OR gate (T_{OR}).

Similarly, for an OR gate, we have

$$s_0 \leq t_{OR} < \frac{s_1}{n_{\max}}$$

Threshold functions in Figs. 5.2 and 5.3 are for $n_{\max} = 8$.

5.3. OTHER BOOLEAN FUNCTIONS

For an inverter, the threshold function is simply, $T_{INV}(V) = 1 - V$. Also,

$$T_{NAND}(V) = 1 - T_{AND}(V)$$

$$T_{NOR}(V) = 1 - T_{OR}(V)$$

Complex Boolean functions can be defined in terms of the threshold functions of primitive gates. For example, the threshold function of an exclusive-OR gate

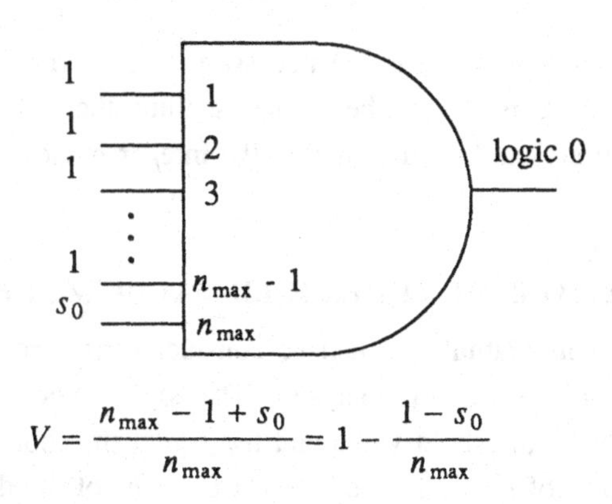

(a) Case 1: logic 0 output with maximum average input value.

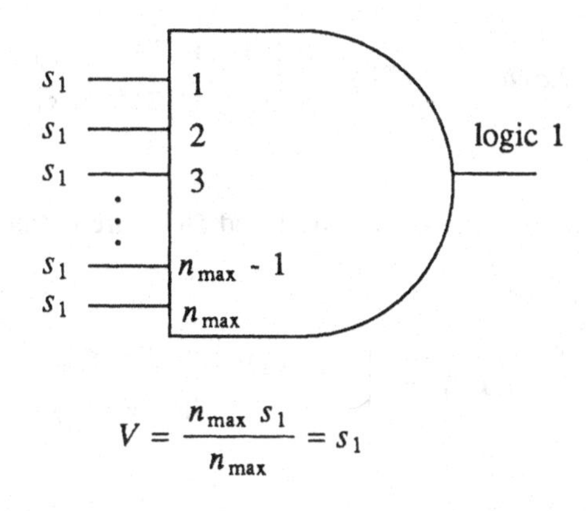

(b) Case 2: logic 1 output with minimum average input value.

Fig. 5.4 Determination of thresholds.

($C = A \cdot \overline{B} + \overline{A} \cdot B$) can be expressed as:

$$V_C = T_{XOR}(V_A, V_B)$$

$$= T_{OR}\left(\frac{T_{AND}\left(\frac{1+V_A-V_B}{2}\right) + T_{AND}\left(\frac{1-V_A+V_B}{2}\right)}{2} \right)$$

where V_A, V_B and V_C are the values of input and output signals of the exclusive-OR gate.

In actual implementation, it is not necessary to perform the threshold-value simulation with floating point numbers. By scaling the value 1.0 to a large integer, all computations can be performed in the integer mode.

5.4. INTERPRETATIONS OF THRESHOLD-VALUE MODEL

The threshold-value simulation can be considered as a combination of logic simulation and dynamic testability analysis†. The logic value of a node is accurately represented by its threshold value and the signal thresholds, s_0 and s_1. The dynamic controllability of the node is expressed in terms of the difference between the thereshold value and the logic value. More specifically, the logic value, as a function of the given threshold value (*TV*), is expressed as follows:

$$LogicValue\,(TV) = \begin{cases} 0 & \text{if } TV \leq s_0 \\ 1 & \text{if } TV \geq s_1 \end{cases}$$

The dynamic 1 and 0 controllabilities *DC* 1 and *DC* 0 are defined as:

$$DC\,1(TV) = \begin{cases} 1.0 - TV & \text{if } TV \leq s_0 \\ 0 & \text{if } TV \geq s_1 \end{cases}$$

$$DC\,0(TV) = \begin{cases} 0 & \text{if } TV \leq s_0 \\ TV - 0.0 & \text{if } TV \geq s_1 \end{cases}$$

If the logic value on a node is 1, i.e., the corresponding *TV* is greater than or equal to s_1, then *DC* 1 is 0. This is because no effort is needed to justify a logic 1 on a

† Testability analysis normally refers to evaluation of controllability and observability of internal signals of a circuit when the primary inputs are not assigned to any specific values. Most computed measures thus represent the effort of controlling or observing the internal nodes. The term *dynamic* is used here to indicate that the testability analysis is being performed for given primary input values. The effort, in this case, is required only to change those signals whose present states are not favorable to the controlling or observing objective.

node if its logic value is already 1. On the other hand, if the logic value is 0, i.e., *TV* is smaller than or equal to s_0, then *DC* 1 is defined as the difference between the strongest 1, i.e., 1.0, and *TV*. The weaker the 0, as indicated by a higher threshold value, smaller is *DC* 1, indicating that the signal is more controllable when a change to 1 is required. The definition of DC0 follows a similar reasoning. By this definition, the relative strength of a signal value is quantitatively expressed.

Notice that threshold functions of primitive gates are piecewise linear and each segment has a different slope. For an AND gate, the slope of the segment with average input value V in the range $[0, t_{AND}]$ is s_0 / t_{AND} which is smaller than the slope in the range $[t_{AND}, 1]$. Therefore, in computing the threshold value of the output of an AND gate, the weight of an input signal with logic 1 is higher than that of an input with logic 0. Similarly, for an OR gate, the weight of a signal with logic 0 is higher than that of an input with logic 1.

The co-domain of the threshold-value function, $[0, s_0] \cup [s_1, 1]$, is smaller than the domain, [0,1]. Thus, the dynamic controllability information is compressed during the evaluation of the gate. In the dynamic controllability determined from the threshold model, the weighting of a signal close to the primary outputs is quite different from that of a signal close to the primary inputs. The dynamic controllability information thus extracted from the threshold value may have no *absolute* meaning. Clearly, it is meaningless to compare the threshold values of two signals in a circuit or in two different circuits. However, if we apply two different input vectors to the *same* circuit, by comparing the threshold values of a specific signal, we can tell which vector is better for driving that signal to a specific logic value. In other words, the threshold-value should be used in a relative sense.

5.5. FANIN-DEPENDENT THRESHOLD-VALUE MODEL

In the model described, the threshold-values are set for the entire circuit based upon the maximum fanin, and therefore, are independent of the number of fanin of an individual gate. In other words, the threshold model of a two-input AND gate is the same as that of a four-input AND gate provided both gates belong to the same circuit. However, the average input value of a two-input gate will only vary in the range $[0, \frac{1+s_0}{2}] \cup [s_1, 1]$, i.e., $[0, 0.55] \cup [0.9, 1]$, while, for a four-input gate, it will vary in the range $[0, \frac{3+s_0}{4}] \cup [s_1, 1]$, i.e., $[0, 0.775] \cup [0.9, 1]$. Therefore, the actual codomain of a two-input gate will be smaller than that of a four-input gate as shown in Fig. 5.5.

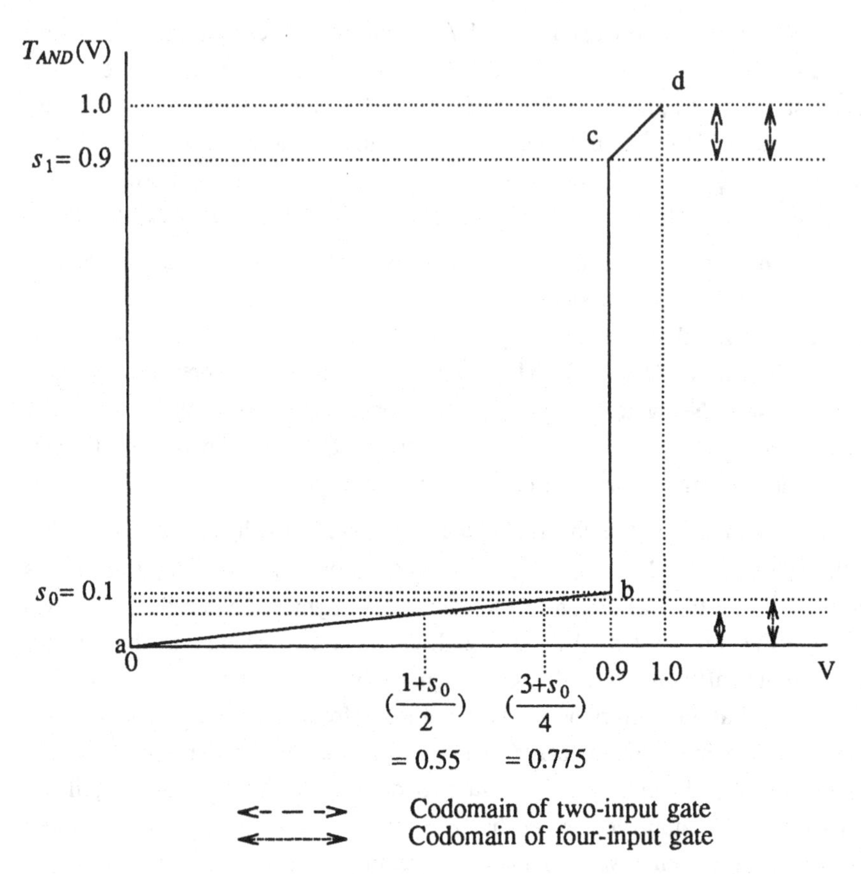

Fig. 5.5 Codomains of logic gates with different fanins.

Another deficiency of the threshold model is the inconsistency between an inverter, a single-input NAND gate, and a single-input NOR gate. For an inverter, the function is $1 - V$ where V is the input value. For single-input NAND and NOR gates, the threshold functions are $1 - T_{AND}(V)$ and $1 - T_{AND}(V)$ as derived from multiple-input gate functions shown in Fig. 5.2 and 5.3. Even though the logical output is correct in all the three cases, the output values will be different for the same input value applied to these implementations of the inverter function.

The two problems just mentioned can be overcome by using a *fanin-dependent* threshold model. In the fanin-dependent model, the threshold functions will be different for gates with different number of fanin signals. Fig. 5.6 shows fanin-dependent threshold functions for AND gates. If the output logic value is 1, the function is represented by the segment $c-d$ which is independent of the

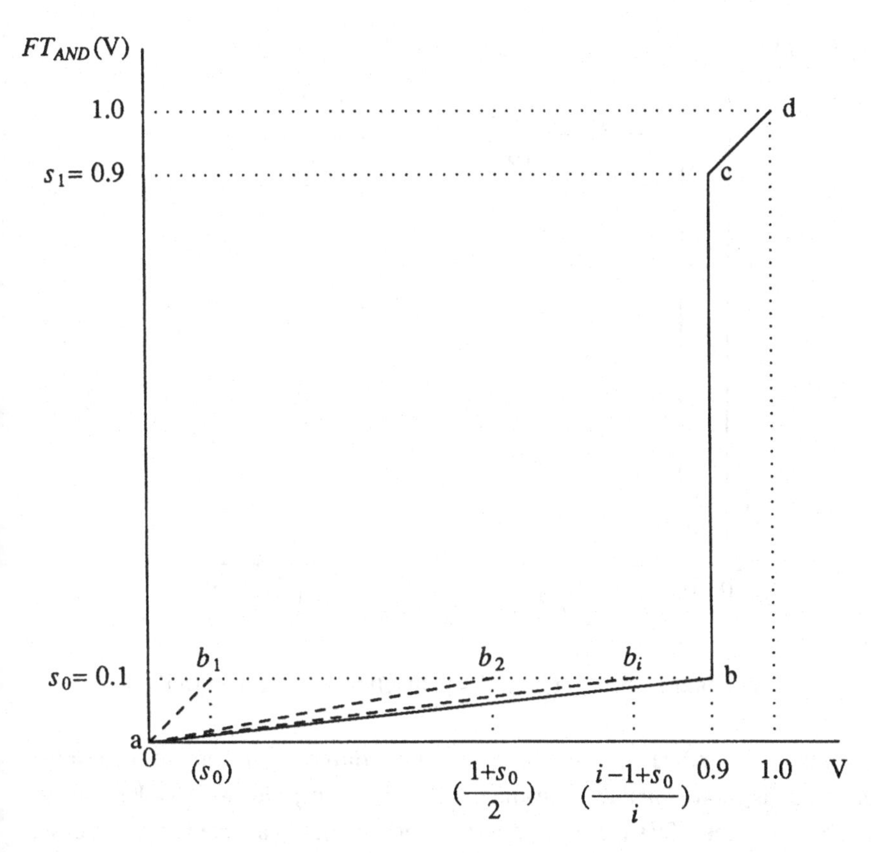

Fig. 5.6 Fanin-Dependent Threshold function for AND gate (FT_{AND}).

number of fanin signals. If the output logic value is 0, it will also be a function of the number of fanin signals. If the number of fanin signals is i (i must be less than or equal to n_{max}, the maximum number of fanin signals for any gate in the circuit), the segment $a-b_i$ is applied. Therefore, the codomain, $[0,s_0] \cup [s_1,1]$, is independent of the number of fanin signals. Similarly, for OR gates, the fanin-dependent model is shown in Fig. 5.7. If the output logic value is 0, the function is represented by the segment $a-b$. If the output logic value is 1, for an i-input gate, the function is represented by the segment c_i-d. In this model, functions of an inverter, a single-input NAND gate, and a single-input NOR gate are identical.

Indeed, the fanin-dependent model is elegant. It, however, has a slightly higher complexity compared to the static model discussed in last section. Some applications of the threshold-value model like the one to test generation as discussed in Chapter 6 need not use the fanin-dependent model. For test generation, threshold values will be used only in a relative sense when the circuit with a *fixed*

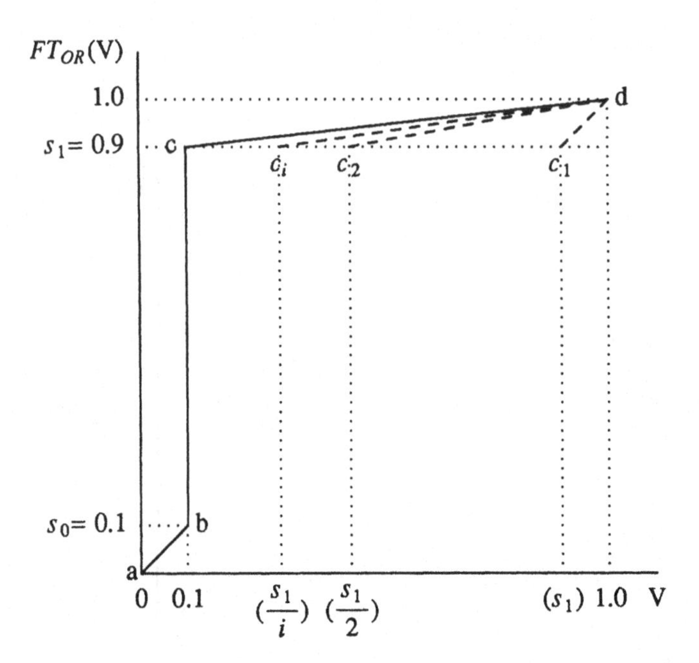

Fig. 5.7 Fanin-Dependent Threshold function for OR gate (FT_{OR}).

implementation is repeatedly simulated through different vectors to determine which vector is more suitable in leading to a test. Thus, the two problems mentioned above do not affect the result and the static model is adopted in the present application.

5.6. THRESHOLD-VALUE SIMULATION WITH THREE STATES

At least three logic values (1, 0, and unknown or X) are needed in both simulation and test generation if there are memory elements in the circuit. In addition, we can devise logic models for buses and tri-state elements to avoid the need for an explicit *floating state* [3]. In this section, we will describe techniques to extend the threshold-value model to handle three signal states.

We consider two ways of extending the threshold-value model introduced in the previous section [4]. We assign a range, say [0.45,0.55], to the unknown state. In the first method, for each type of logic gate, a set of three threshold models are defined. Each threshold model only handles two signal states. Thus, the three models are defined for pairs of input values, (0,1), (0, X), and (X,1), respectively. For example, Fig. 5.8 shows the threshold function $T_{AND_{x1}}$ for an AND gate with input states X and 1.

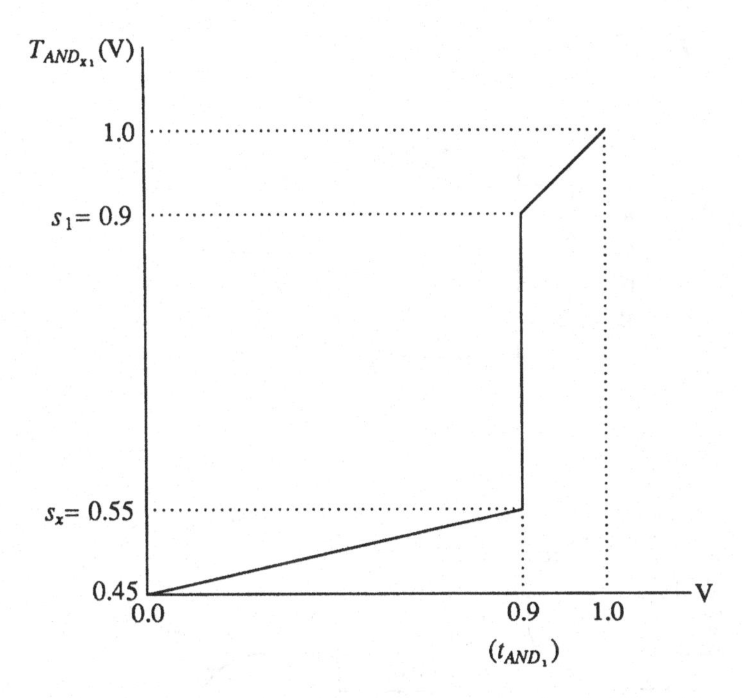

Fig. 5.8 Threshold function with input state *X* and 1 for AND gate.

Since the threshold model discriminates only between two states, the evaluation of a gate using three signal states requires two steps. For example, consider an AND gate with five inputs 10XX1 as shown in Fig. 5.9. We split the gate into two AND gates. The first gate has three inputs with values 0XX. The output of this gate is computed from a three-input threshold model. The resulting output is applied to a second three input AND gate whose other inputs are 11. Thus, by a dynamic two-level splitting, any gate can be evaluated. One problem with this method is that different ways of splitting will result in slightly different output values. Note, however, that the logic value of the output will always be correct. Since the threshold value is used only in a relative sense, that is, only to compare two vectors, the implication on test generation will not be serious.

In the second method, all three logic states are simultaneously considered. Since it is impossible to avoid ambiguity if the gate output is evaluated only from the average of input values, we determine the logic value of the output separately. For example, for a three input AND gate, the average input values are the same for inputs *XXX* and 1*X*0. Both are 0.5. The logic output is, however, different in the two cases. For each gate type, two separate threshold functions are defined corresponding to the output logic state. Once the output logic value is determined,

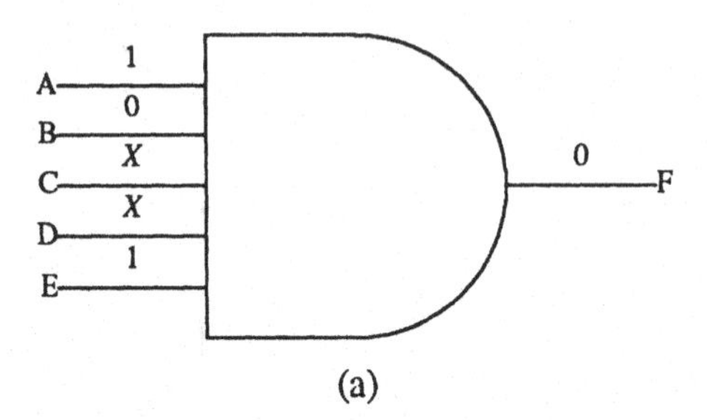

(a)

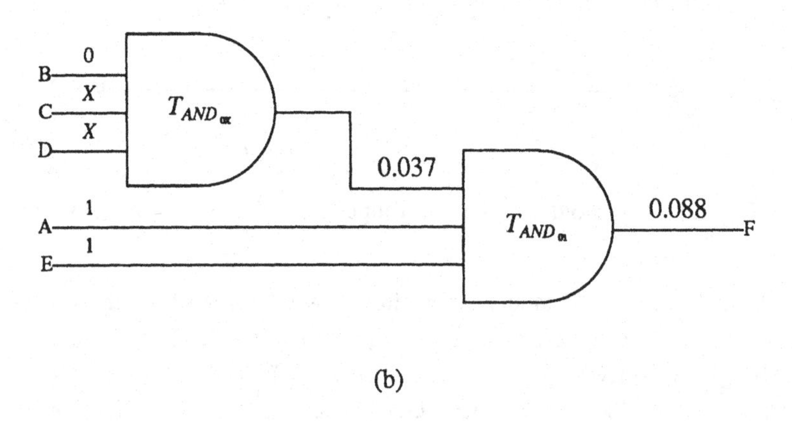

(b)

Fig. 5.9 Dynamic two-level splitting for gate evaluation.

the appropriate function is applied for computing the threshold value. Figure 5.10 shows this threshold model for an AND gate. The function represented by a-b-d is applied if the output logic value is 0. Otherwise, the function c-e-f-g is used. In the example cited above, the input average value was 0.5 for inputs *XXX* and 1*X*0 and the logic values of the output are *X* and 0, respectively. The function c-e-f-g is used to evaluate the output threshold value for the first case, while the function a-b-d is used for the latter case.

In the program, TVSET, described in the next chapter, the second method is adopted because of its lower computation overhead.

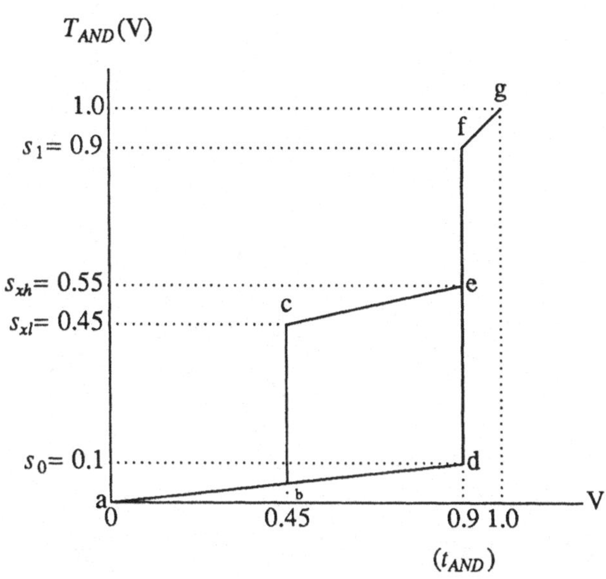

Fig. 5.10 Threshold function for AND gate with three input states.

REFERENCES

[1] K. T. Cheng and V. D. Agrawal, "A Simulation-Based Directed-Search Method for Test Generation," *Proc. Int. Conf. Computer Design. (ICCD'87)*, Port Chester, NY, pp. 48-51, October 1987.

[2] V. D. Agrawal and K. T. Cheng, "Threshold-Value Simulation and Test Generation," *Testing & Diagnosis of VLSI & ULSI (Proc. NATO Adv. Study Inst. Como, Italy, June 1987)*, M. Sami and F. Lombardi, Editor, Kluwer Academic Publishers, Dordrecht, The Netherlands, 1988.

[3] V. D. Agrawal, S. K. Jain, and D. M. Singer, "Automation in Design for Testability," *Proc. Custom Integrated Circuits Conf.*, Rochester, NY, pp. 159-163, May 1984.

[4] K. T. Cheng, V. D. Agrawal, and E. S. Kuh, "A Sequential Circuit Test Generator Using Threshold-Value Simulation," *18th Fault-Tolerant Computing Symp. (FTCS-18) Digest of Papers*, Tokyo, Japan, pp. 24-29, June 1988.

Chapter 6

TEST GENERATION USING THRESHOLD-VALUE SIMULATION

Having defined the threshold model, we now proceed with test generation which is the central topic of this book. Our approach is based on simulation. We use serial fault simulation with the static threshold-value model.

6.1. FAULT ANALYSIS USING THRESHOLD-VALUE MODEL

To simulate the effect of a stuck fault, the value of the faulty line is simply forced to the faulty value (0.0 for a stuck-at-0 fault or 1.0 for a stuck-at-1 fault.) The difference in values between the outputs of the good circuit and the faulty circuit gives an indication of not only whether the fault is detected but also how far the fault is from being detected.

As an illustration, consider the circuit in Fig. 6.1. The result of simulation with an input vector 100 is shown. The values in parenthesis correspond to the faulty circuit with line E stuck-at-0. Notice that the good circuit output 0.002 corresponds to a logical 0 and so does the faulty output 0.0. Thus the fault is not detected. However, the difference between the two values is due to the fault. A

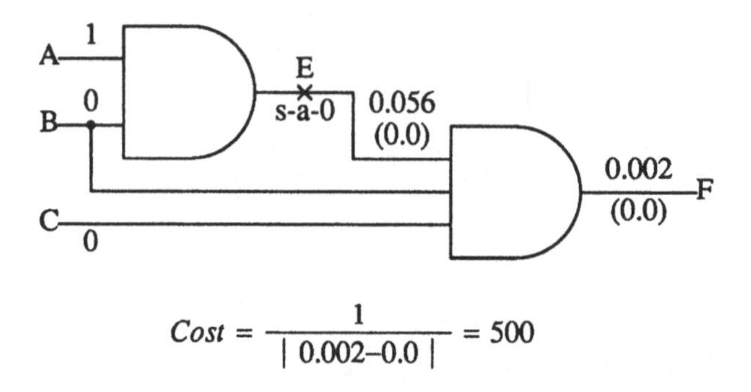

$$Cost = \frac{1}{|\,0.002 - 0.0\,|} = 500$$

Fig. 6.1 Fault simulation example.

difference greater than the *gap* (0.9 – 0.1 = 0.8) between the logical 1 and 0 values, as shown in Figs. 5.2 and 5.3, would indicate detection of the fault.

We evaluate this pattern for its closeness to a test of the given fault by computing a *cost*. As shown in Fig. 6.1, the cost is defined as reciprocal of the absolute difference between the good and the faulty circuit output values. For multiple output circuits, the cost of the vector is defined as the minimum cost among the primary outputs, i.e.,

$$Cost = \min_{i \in PO} \frac{1}{|\,TV_g(i) - TV_f(i)\,|}$$

where PO is the set of primary outputs and $TV_g(i)$ and $TV_f(i)$ are the threshold values of signal i in fault-free and faulty circuits, respectively. Obviously, an *infinite* cost will indicate no relationship to a test while a cost less than $1/0.8 = 1.25$ indicates that the input vector is a test.

6.2. TEST GENERATION

Using the above cost function, we can perform a directed search for a test for the given fault. We begin with any vector and compute the cost for this vector and for all its neighbors defined as vectors at unit Hamming distance. Ideally, the neighbor that has the lowest cost, becomes the next vector in the search process. In practice, however, for any arbitrary order in which the neighboring vectors are processed, one can adopt a *greedy* heuristic and accept the first one that lowers the cost.

6.2.1 A Simple Example

Fig. 6.2 illustrates the above method. Assume that the starting vector is 100 and as shown in Fig. 6.1 the cost is 500. A one-bit change on the input A produces an infinite cost and hence is not acceptable (Fig. 6.2a). We, therefore, try a change in the input B. As shown in Fig. 6.2b, this reduces the cost to 27.02. Using the greedy heuristic, we accept this change. Our new vector is 110 with a cost of 27.02. Next, a one-bit change in C leads to fault detection (Fig. 6.2c).

6.2.2. A Multipass Process

It is well known that a cost minimization procedure can, at times, lead to local cost minimum. If that happens, one restarts at a different point in the search space. In the above example, we were fortunate. A test generator using the directed search principle, in general, requires multiple passes. To illustrate this, we will first describe the combinational test generator [1] and then discuss the differences required in handling sequential circuits [2].

First Pass. For combinational circuits, the first pass is executed as shown in Fig. 6.3. All faults under consideration are simulated through a threshold-value fault simulator. Cost of the initial vector is computed for each fault. The fault with the lowest cost is selected as the target fault for test generation. A directed search for test is conducted. We use a variable "#trials" to denote the number of trial vectors generated in the search process for the current target fault and an integer "n" to denote the bit position in the vector that is changed. The "SearchLimit" is a parameter denoting the limit on the number of trials in the search. Before applying a trial vector, the signal states are saved to allow a return to the original state if the trial vector does not produce lower cost. If the trial cost is lower than the current cost but higher than the cost threshold C_0 that is defined as $1/(s_1 - s_0)$, the trial vector is accepted as the current vector but the search continues. If a test is found, all faults are simulated with this vector and the detected faults are removed from consideration. As a result of this threshold-value fault simulation, cost of this test vector is also obtained for each undetected fault. Again, the fault with the lowest cost becomes the target for the next test.

If the search for a test for some fault terminates without the fault being detected, then that fault is considered as *hard to detect*. This fault is not classified as detected or redundant but it is excluded from test generation again in the current pass. It can, however, be detected by other vectors through fault simulation.

Subsequent Passes. There are two possible reasons for a fault not being detected during a pass: (1) the fault is redundant, or (2) the search terminates at a *local*

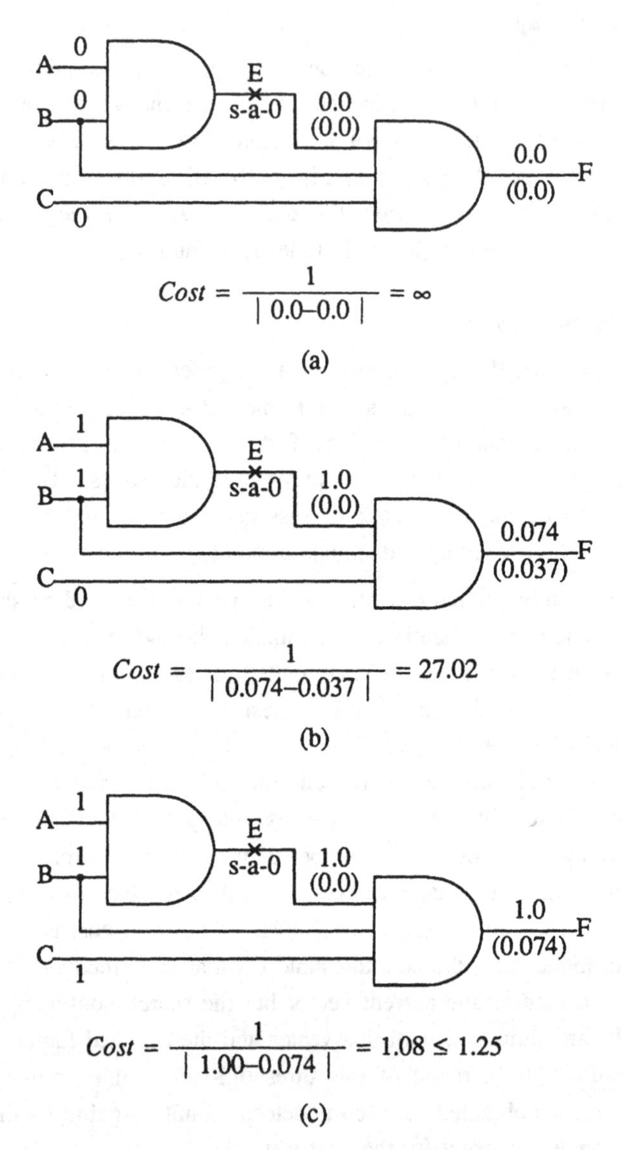

Fig. 6.2 Illustration of test generation.

minimum in the search space. In the latter case, the often used procedure is to start with a different initial vector. This is exactly what the subsequent passes do. The algorithm is the same as that used in the first pass except that the starting fault list for a subsequent pass is the *hard to detect* list generated by the previous pass. Any number of passes can be made at user's option. Each pass can be initiated either with a new random vector or with a vector that is at a large Hamming

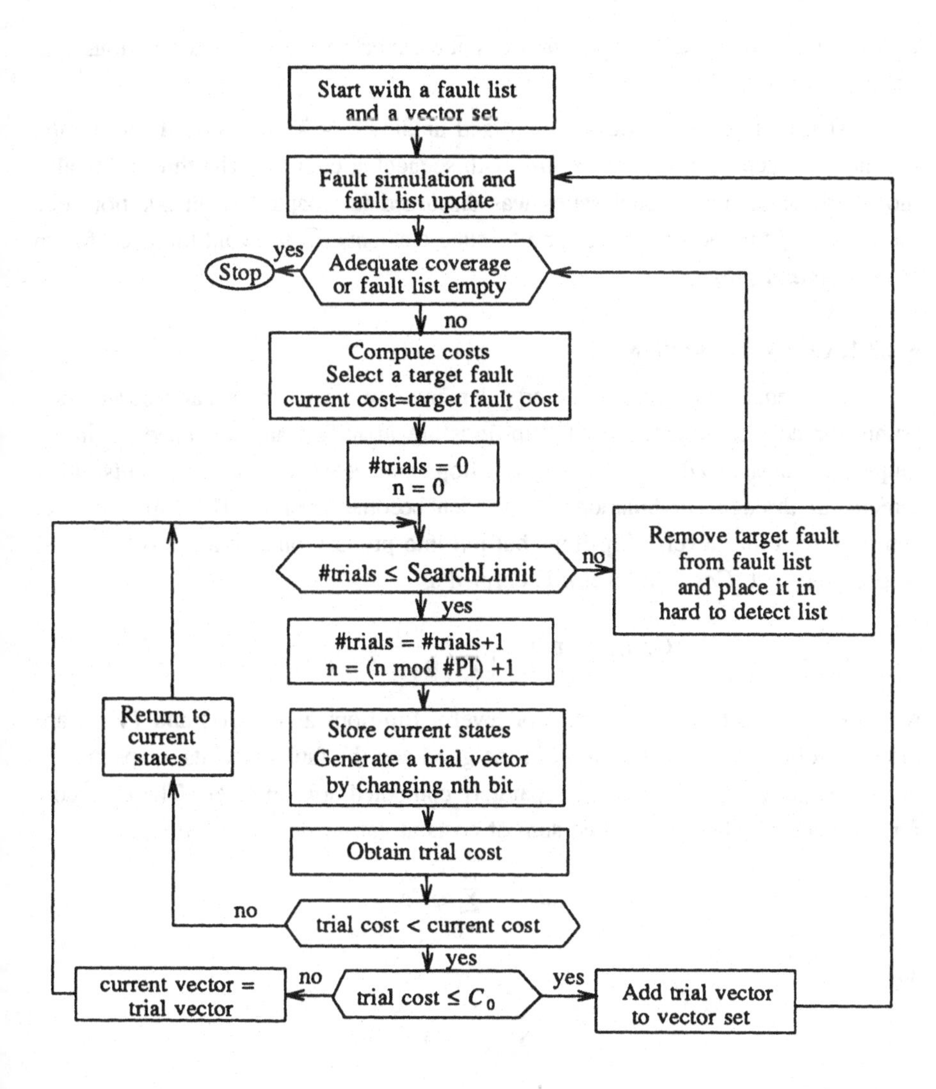

Fig. 6.3 Test generation through directed search.

distance from the vectors generated in all previous passes.

6.2.3. Requirements of Sequential Test Generator

An automatic test generator for sequential circuits must initialize the circuit and generate race-free and hazard-free tests for modeled faults. It should also handle, in general, both synchronous and asynchronous circuits since even the "so called" synchronous designs often use some asynchronous logic for various reasons.

At least three logic values (1, 0, and unknown or X) are needed for simulation and test generation when dealing with sequential circuits. The threshold-value model for these three signal states was discussed in Chapter 5. In addition, one must use logic models for buses and tri-state elements [3] to avoid the need for an explicit *floating state*.

6.2.3.1. Cost Computation

In a sequential circuit, memory elements are treated as pseudo-observation points for cost computation. All flip-flops are identified and levelized. Primary outputs are designated level 0. The flip-flops connected to primary outputs either directly or through combinational logic then become level 1. The flip-flops that can only feed into level 1 flip-flops but not into primary outputs are level 2. And so on. The cost, $Cost_n$, at level n is defined as

$$Cost_n = \min_{i \in I_n} \frac{1}{|TV_g(i) - TV_f(i)|}$$

where I_n is the set of input signals of level n flip-flops and $TV_g(i)$ and $TV_f(i)$ are the threshold values of the signal i in the good and the faulty circuits, respectively. For an input vector the costs are separately computed at each level. The total cost for this vector is then a weighted sum of costs at various levels. That is,

$$Cost = \sum_{n=0}^{m} a_n Cost_n$$

and

$$\sum_{n=0}^{m} a_n = 1$$

where m is the maximum level in the circuit. The values of a_n are implementation parameters. A suggested way of computing these, also used in the TVSET program described later, is as follows:

$$a_n = \frac{10^{m-n}}{\sum_{k=0}^{m} 10^k}.$$

For example, if the maximum level in the circuit is 2, $a_0 = 100/111$, $a_1 = 10/111$, and $a_2 = 1/111$. If the fault is not detected at the level 0 (primary output), i.e., $Cost_0 > \frac{1}{s_1 - s_0} = 1.25$, then the weighted cost is used to direct the search.

Potential detection can be used as an option in test generation. If this option is selected, the fault is treated as potentially detectable whenever $1/s_{xh} \leq Cost_0 \leq 1/(s_{xl} - s_0)$. That is, for at least one primary output, the logic values of the good circuit and the faulty circuit differ but one of them is an unknown state.

The fault effects are propagated through gates and flip-flops, level by level, by gradually minimizing the cost function. To illustrate how cost function helps the search, we give an example in Table 6.1 showing typical cost variations during test generation. There are two levels of flip-flops in the circuit. The fault effect, that appears at level 2 after vector $V2$ is simulated ($Cost_2 \leq 1.25$), is propagated to level 1 after $V4$ is simulated. Finally, the fault effect is propagated to the primary output (level 0) after $V6$ is applied. The vector sequence $V1$ through $V6$ is the generated test in this case.

Table 6.1 Cost Variation during Test Generation

Vector	$Cost_0$	$Cost_1$	$Cost_2$	Comments
$V1$	∞	∞	27.5	
$V2$	∞	∞	1.1	Fault effect appears in level 2
$V3$	∞	14.3	1.1	$Cost_1$ reduced
$V4$	∞	1.2	12.3	Fault effect appears in level 1
$V5$	35.2	1.2	12.1	$Cost_0$ reduced
$V6$	1.1	1.2	12.1	Fault detected

6.2.3.2. Initialization

Test generation can be started with all unknown states in both good and faulty circuits. The test generator then automatically initializes the circuit. Alternatively, a user-supplied initialization sequence can be used. All faults are simulated with this sequence to find the lowest cost fault as the target for test search.

The fault simulation also yields the good and the faulty states that are used at the beginning of the search. After a test sequence is generated, subsequent fault simulation, in a similar manner, provides the states for the next target fault.

6.2.3.3. Race Analysis

We use a logic model proposed by Ulrich and Baker [4] for automatic race analysis. This model requires one additional AND gate per cross-coupled NOR latch and one additional OR gate per cross-coupled NAND latch. For example, Fig. 6.4 shows the model for a cross-coupled NOR latch. The AND gate has an input permanently set to the unknown state (*X*). Whenever an input sequence, that can cause a race (11 followed by 00, 1*X* followed by 00, or *X*1 followed by 00), is applied to the latch, the output is automatically set to the unknown state to settle the race condition. Although the gates can have any arbitrary rise and fall delays, in this example they are assumed to have one unit of rise/fall delay as indicated by 1/1.

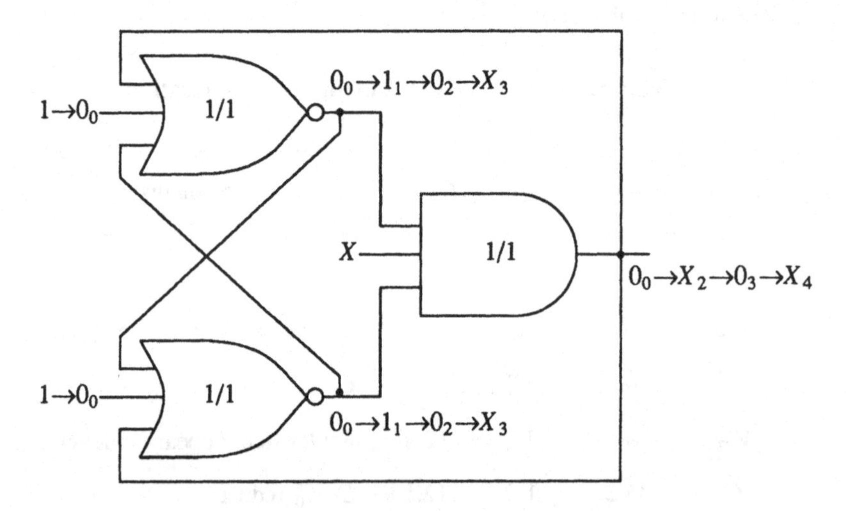

Fig. 6.4 Model for automatic race analysis.

6.2.3.4. Feedback Loop Analysis

In the case of a feedback, even when the logic values are stable, the continuous values used in the threshold-value model may require repeated evaluation of the gates that form the loop. It is, however, possible to avoid such iterative evaluation by identifying loops prior to simulation and using closed form expressions to compute feedback signals. As an example, Figure 6.5 shows a cross-

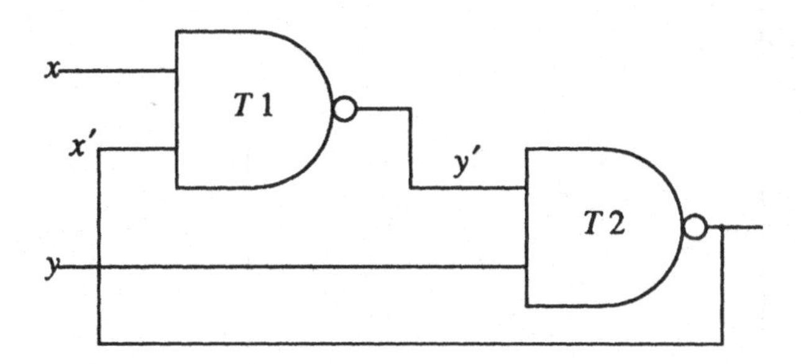

Fig. 6.5 Feedback loop analysis for cross couple NAND latch.

coupled NAND latch. When signal values x and y are applied to the inputs of the latch, the final stable threshold value x' and y' can be determined by solving

$$x' = T_{NAND}\left(\frac{y + T_{NAND}(\frac{x + x'}{2})}{2}\right).$$

Since $T_{NAND}(V)$ is a piecewise linear function of V, The equation can be solved for combinations of logic values of x and y. Thus, if x and y both lie in the range of logic 0 ([0,0.1]), then line d-c in Fig. 6.6 will be used for evaluating both gates $T1$ and $T2$. The above equation then becomes

$$x' = 1 - \frac{1}{18}\left(1 - \frac{x + x'}{18} + y\right)$$

from which x' can be easily computed. Similarly, all possible cases for this latch can be analyzed in advance. During simulation, the output values of latches are then computed directly without iterations.

Since the problem of finding all feedback loops, in general, is NP-complete, only the feedback loops that consist of two gates (e.g., cross-coupled latches) may be identified and analyzed in this manner. For any other loops, the simulator will continue the event-driven evaluation process until the values converge. It may, however, assume an oscillation if the number of such evaluations exceeds certain prespecified limit in which case all active gates could be set to the unknown state.

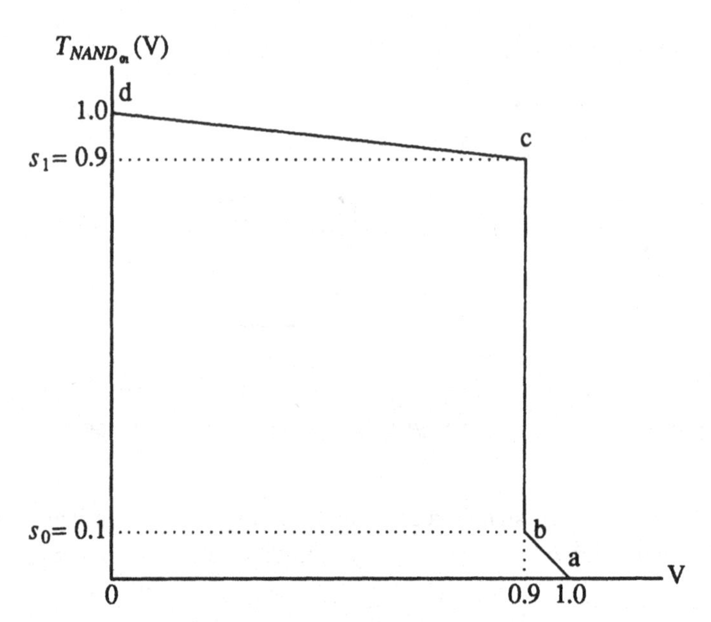

Fig. 6.6 Threshold-value model of NAND gate ($T_{NAND_\alpha}(V)$).

6.2.3.5. Switching the Target Fault

The test generator produces vector sequences for one target fault at a time. The target fault is selected from the fault list as the one having the lowest cost. During the search process, that is guided by the cost of the target fault, it is quite possible that the cost of some other fault becomes lower than that of the target fault. This indicates that the current state is more favorable for testing the other fault. Target switching in such cases is found to be a good strategy. Since fault simulation is performed on all undetected faults whenever a new vector is generated, all corresponding states are readily available and their costs can be computed without much overhead. Similar benefits of switching targets in combinational test generation have been reported by other authors [5].

6.2.3.6. Synchronous and Asynchronous Modes

Test generation can be performed in any one of the two modes described below depending upon the circuit structure and its operation.

Synchronous Circuits. In synchronous circuits, the clock signals are identified and their sequences are specified to the test generator. The flow chart of test generation for synchronous sequential circuits is shown in Fig. 6.7. The non-clock (or data) signals are manipulated by the test generator in a manner similar to that for combinational circuits. That is, one bit is flipped to obtain a trial vector for

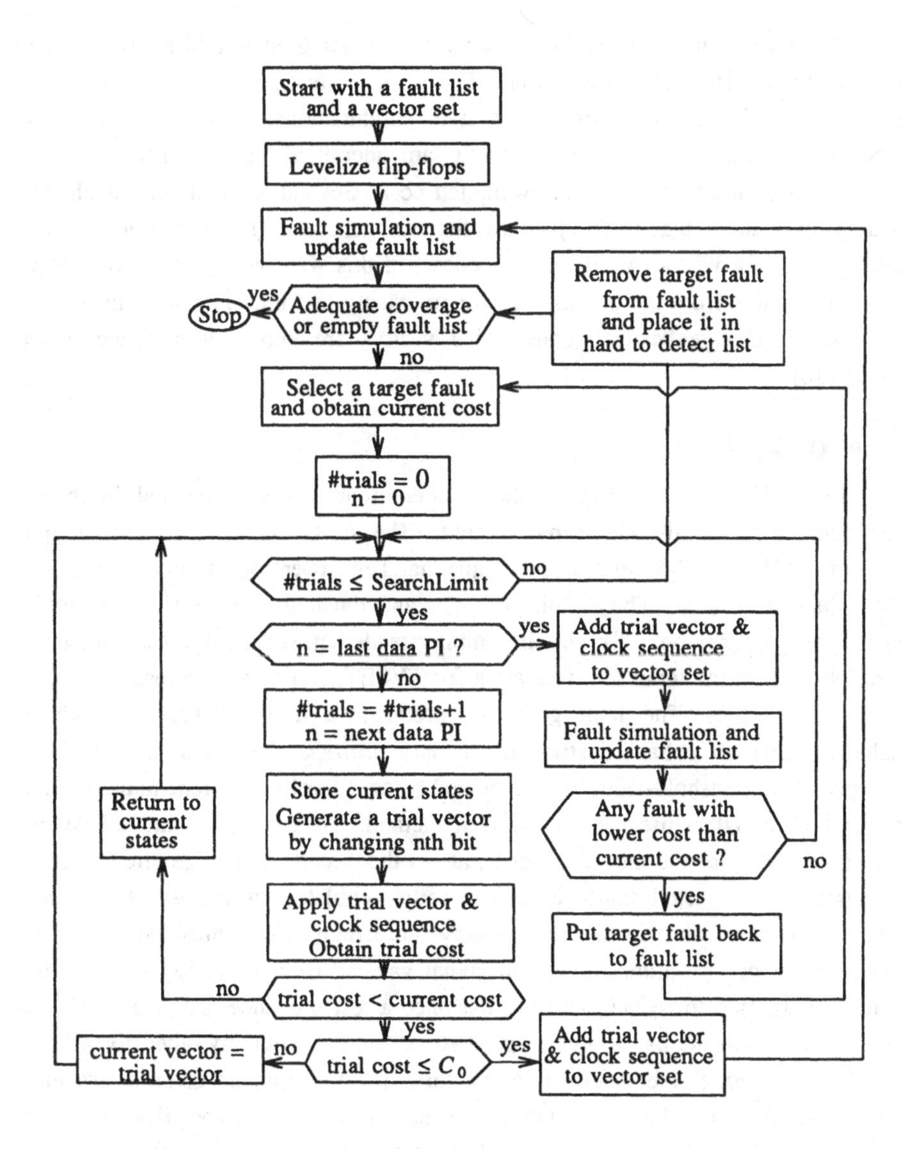

Fig. 6.7 Test generation for synchronous sequential circuits

which cost is computed. The change is accepted whenever the new cost is lower. After all data signal bits have been processed once, only the final vector is added to the test sequence. This vector is then followed by the prespecified clock sequence. Cost is again computed. The search uses combinational mode and clock sequence, repeatedly, until the fault is detected or the search for the present target is abandoned due to a local cost minimum.

Asynchronous circuits. When no clock signal is identified all signals are manipulated as data during the search. However, the test generation for asynchronous sequential circuits differs from that of combinational circuits in two ways. First, the outputs of memory elements are identified and used as pseudo-observation points to compute the weighted cost. Second, each single bit change at a primary input that is accepted by the test generator (because it lowers the cost) produces a vector in the test sequence. In this way, any new vector that is generated differs from the previous vector in exactly one bit. Test sequences with just one bit change between consecutive vectors are known to produce fewer hazards [6].

6.3. PROGRAM

Fig. 6.8 shows a test generation system that was implemented in the C language for use in the Unix environment. The system consists of three major programs: TVSET (Threshold-Value SEquential Test generator); a circuit compiler; and a fault generator. The circuit compiler and the fault generator accept logic level circuit description. The circuit compiler reads netlist (gate-level connectivity description) from an input file, formats it for TVSET, and stores the result in a circuit database file. The fault generator produces a fault database file which includes either all collapsed stuck faults or a user-specified fault list. TVSET incorporates a threshold-value simulator and a logic-level concurrent fault simulator. Both simulators use the event-driven method. Arbitrary rise and fall delays can be specified for the output of each gate in the circuit. For large circuits, only a subset of undetected faults is concurrently simulated in the threshold-value mode. This is because in this mode many more fault effects must be processed due to the wider variation allowed in signal values. Concurrent logic-level fault simulation is used to reduce the fault list once a test sequence is found. TVSET begins either with the list of all collapsed stuck faults or with any other given fault list. It can accept a given set of initial vectors. If no vectors are given, it assumes the circuit initially to be in the unknown state. The user also specifies the mode (combinational, synchronous sequential or asynchronous sequential), the clock sequence if synchronous sequential mode is selected, and, if desired, the "no potential detection" option (the default assumes potential detection).

6.4. EXPERIMENTAL RESULTS

In the following, we will discuss the results on combinational and sequential circuits.

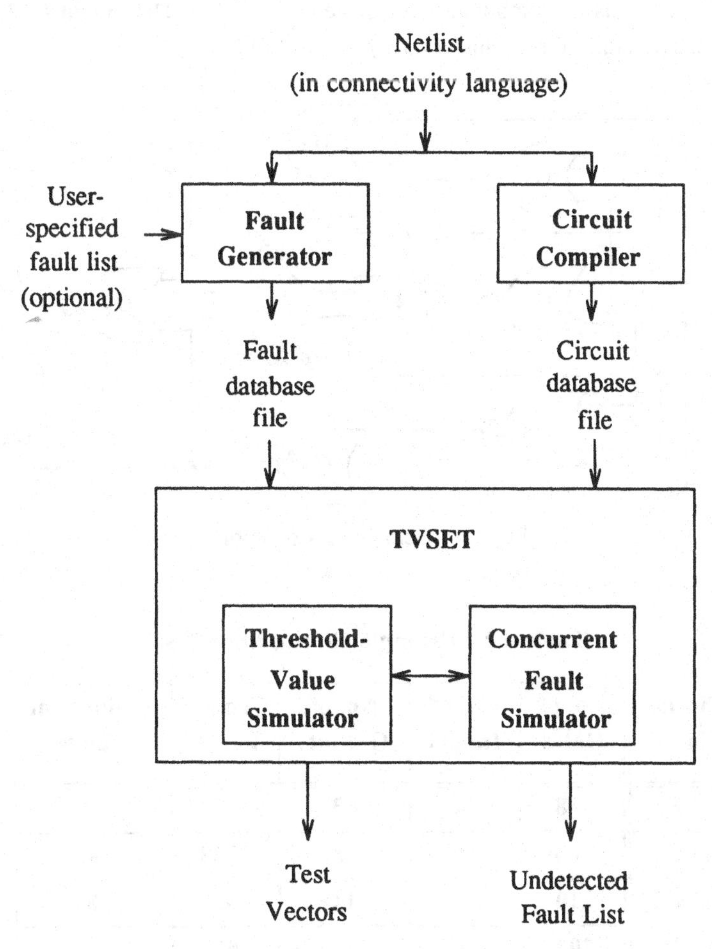

Fig. 6.8 The threshold-value test generation system

6.4.1. Combinational Circuits

Results on seven combinational circuits are presented. Table 6.2 gives their statistics. The first circuit (S. E.) is Schneider's example [7] that is often used to check test generation algorithms for their capability to sensitize multiple paths. The circuit diagram is shown in Fig. 6.9. A feature of this circuit is that a test for G_2 output stuck-at-0 fault can not be generated by a single-path sensitization method that attempts to sensitize only one path at a time. This fault, however, is testable. The two ALUs are the Texas Instruments' four-bit ALU circuit 74181 [8] and its eight-bit ripple-carry extension. The exclusive-OR gates in the ALU's were replaced by an equivalent five-gate AND, OR, NOT implementation. Last three circuits, C432, C499, C880, and C1355, are taken from the benchmarks used in recent evaluations of combinational test generators [9]. The circuit C1355 was obtained by expanding the exclusive-OR gates in C499.

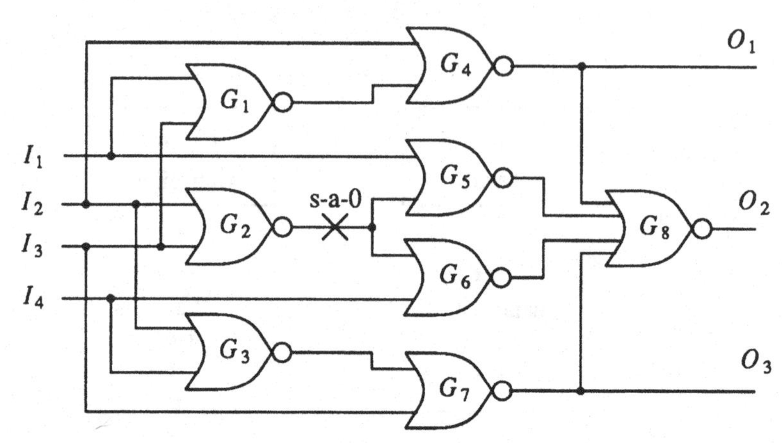

Fig. 6.9 Schneider's example.

Table 6.2 Example Circuit Statistics

Circuit Name	No. of Gates	No. of Inputs	No. of Outputs	Total Faults	Redundant Faults
S. E.	8	4	3	34	2
ALU4	104	14	8	268	4
ALU8	210	22	11	524	8
C432	203	36	7	524	-
C499	275	41	32	758	-
C880	469	60	26	942	0
C1355	619	41	32	1574	9

The result of two passes of TVSET is shown in Tables 6.3 and 6.4. For Schneider's example, all except the two redundant faults were detected in the first pass. In the directed search approach, a fault may not be detected because either it is redundant or the search stopped due to a local cost minimum. There is no way to distinguish between these two situations. The number of redundant faults in Table 6.2 was determined by another exhaustive algorithm [10]. Since our method

Table 6.3 Results on Combinational Circuits After Pass One

Circuit	Pass 1			
Name	Fault coverage %	# of tests	CPU sec. for test gen. VAX11/780	Average vectors simulated per fault
S. E.	94.1	9	0.04	3
ALU4	97.4	33	7.05	12
ALU8	97.9	63	27.04	20
C432	97.7	56	67.22	24
C499	98.3	65	152.05	33
C880	99.2	116	186.50	39
C1355	98.2	107	1165.18	36

Table 6.4 Results on Combinational Circuits After Two Passes

Circuit	After Two Passes			
Name	Fault coverage %	# of tests	CPU sec. for test gen. VAX11/780	Av. vectors simulated per fault
S. E.	-	-	-	-
ALU4	98.5	36	9.09	13
ALU8	98.5	65	36.42	21
C432	98.7	61	84.12	26
C499	98.9	70	208.30	35
C880	99.7	121	232.35	40
C1355	99.1	116	1492.62	38

is based on simulation it considers propagation of fault effects through all paths from the fault site to outputs, detection of faults requiring sensitization of multiple paths presents no problem. On all circuits, the first pass produced a reasonably high fault coverage. The "average vectors simulated per fault" is the average number of one-bit changes (counting even those changes that did not reduce cost) attempted per test. A second pass on the ALU circuits detected all but the redundant faults. For the four benchmark circuits, the fault coverage could be further increased by subsequent passes.

CPU times in Tables 6.3 and 6.4 are for VAX-11/780 and do not include the time of fault simulation over all remaining faults after a test was found. Thus, considering test generation alone, this time may be comparable to a PODEM program [11]. The CPU time, however, is implementation dependent. We feel it can be significantly reduced if fault simulation during test generation is done concurrently. Further improvement is possible through concurrent simulation of several one-bit input changes. The main advantage of the present method, as we will show in the next section, is its applicability to sequential circuits.

It is interesting to note that the average number of vectors simulated per test is on the order of the number of primary inputs. Thus the empirical complexity of our test generation method appears to be proportional to

$$(\textit{Number of Gates}) \times (\textit{Number of Primary Inputs}).$$

We assume that simulation complexity per vector is proportional to the number of gates in the circuit. This is because we stop at the nearest cost minimum instead of guaranteeing a test; guaranteed test generation still remains *NP-complete* [12].

6.4.2. Sequential Circuits

We give results for several sequential circuits. The circuit characteristics are listed in Table 6.5. Only one circuit, MANNY*, is completely asynchronous. Fig. 6.10 shows a schematic of this circuit. SSE and PLANET are finite-state machines with PLA-like implementation†. MULT4 is a 4-bit Booth multiplier designed by an automatic synthesis program. TLC is a traffic light controller circuit. CHIP-A is a CMOS chip designed for a graphics terminal. All circuits, except MANNY, are completely synchronous.

Table 6.6 shows the results obtained from TVSET. For comparison, the results of another sequential circuit test generation program, STG [13], are also

* Courtesy of P. Goel of Gateway Design Automation Corp.

† Obtained from Tony Ma of University of California, Berkeley, CA.

Table 6.5 Sequential Circuits used for Test Generation

Circuit	# Gate	# Input	# Output	# FFs	# Faults
MANNY	26	5	3	7	67
SSE	207	8	7	6	454
MULT4	382	10	9	15	540
TLC	355	4	6	21	772
PLANET	690	9	19	6	1582
CHIP-A	1112	13	13	39	1643

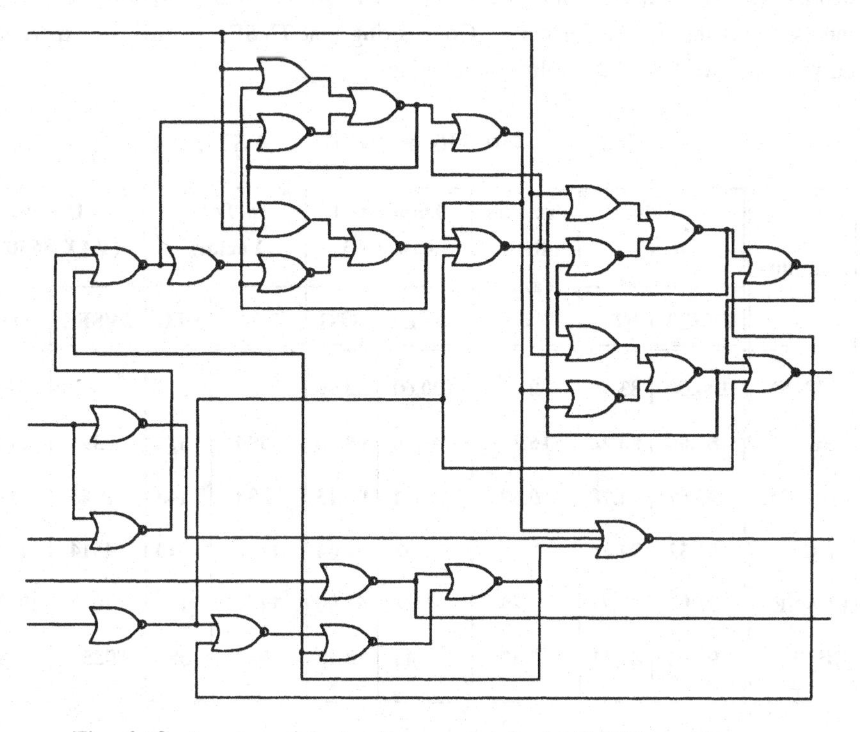

Fig. 6.10 A sequential circuit example, the MANNY circuit.

given. STG uses the conventional method of time-frame expansion and applies the D-algorithm in a backward time processing mode. The performance of TVSET is better than STG in fault coverage and CPU time. In the case of STG, the backtracing in the time-frame expanded circuit results in the high complexity of test generation. Signal values set during backtrace are checked for consistency. The process must backtrack whenever an inconsistency is found which frequently happens when the circuit has reconvergent fanouts. Experience with STG has shown that the lower fault coverage is often due to a practical limit on CPU time and due to *bad* tests that cause races and hazards. The fault coverage, however, increases only marginally if more CPU time is allowed. In our approach, simulation always processes the circuit from inputs to outputs without backtracing and with automatic analysis of races and oscillations. The number of vectors simulated for finding a test vector is related to the number of primary inputs. Assuming that the simulation complexity of the event-driven method for each trial vector with one-bit change is proportional to the number of gates in the circuit, the total complexity of TVSET would be the test length, times the number of primary inputs, times the number of gates. Considering that TVSET is only an experimental program, we find the results encouraging.

Table 6.6 Results of TVSET and STG

Circuit	Fault Cov. (%)		Provable Red. Faults (%)	Total Fault Cov. (%)		# Test Vectors		CPU Secs. (VAX 8650)	
	TVSET	STG		TVSET	STG	TVSET	STG	TVSET	STG
MANNY	100.00	83.95	0	100.00	83.95	35	219	5	NA‡
SSE	81.06	83.20	16.30	97.36	99.50	590	676	398	1134
MULT4	96.67	92.78	0.37	97.04	93.15	388	148	1068	1490
TLC	93.52	94.64	NA	93.52	94.64	1256	5340	4514	32590
PLANET	92.92	57.71	3.29	96.21	61.00	1469	132	4621	19388
CHIP-A	93.06	84.11	4.37	97.43	88.48	924	384	89351	NA

‡ NA ≡ Not Available at the time of writing.

Fault coverage data reported in Table 6.6 were verified through an MOS fault simulator. None of the test generation programs used were capable of identifying redundant faults. The redundancies of 16.3% and 3.29% in SSE and PLANET, respectively, are typical of PLA structures [14]. Redundant faults in MULT4 and CHIP-A were manually analyzed and were found to be undetectable due to the specific logic model used for CMOS tristate devices. STG does not have an interactively running fault simulator. Thus, the fault list is not updated and the test generator separately produces vectors for each fault. Also, for each test, a new initialization is attempted. These two reasons add to the increase in the run time of STG. STG's tests for MANNY had problems because of the asynchronous operation in the circuit.

The number of test vector generated by TVSET, in several cases, is on the same order as that of the other method. This is an indication that the cost function can efficiently direct the search in the input vector space. The high fault coverage achieved by TVSET indicates that the particular cost function is well-suited for avoiding local cost minima.

TVSET was further compared with a commercially available test generation program [15]. This program uses an extended backtrace technique for an efficient implementation of the path sensitization procedure. It generated 330 vectors for the CHIP-A circuit. Although it reported a higher coverage, the same fault simulator that was used for the results of Table 6.6 gave a coverage of 90.75% for these vectors. This is not surprising since results from different simulators often vary due to modeling differences. Considering the 4.37% redundant faults in this circuit the fault coverage can be updated to 95.12%. In comparison, the coverage obtained by TVSET is 97.43% and that by STG is 88.48%.

REFERENCES

[1] K. T. Cheng and V. D. Agrawal, "A Simulation-Based Directed-Search Method for Test Generation," *Proc. Int. Conf. Computer Design. (ICCD'87)*, Port Chester, NY, pp. 48-51, October 1987.

[2] K. T. Cheng, V. D. Agrawal, and E. S. Kuh, "A Sequential Circuit Test Generator Using Threshold-Value Simulation," *18th Fault-Tolerant Computing Symp. (FTCS-18) Digest of Papers*, Tokyo, Japan, pp. 24-29, June 1988.

[3] V. D. Agrawal, S. K. Jain, and D. M. Singer, "Automation in Design for Testability," *Proc. Custom Integrated Circuits Conf.*, Rochester, NY, pp. 159-163, May 1984.

[4] E. G. Ulrich and T. Baker, "Concurrent Simulation of Nearly Identical

Digital Networks,'' *Computer*, Vol. 7, pp. 39-44, April 1974.

[5] Y. Takamatsu and K. Kinoshita, ''CONT: A Concurrent Test Generation Algorithm,'' *Fault-Tolerant Computing Symp. (FTCS-17) Digest of Papers*, Pittsburgh, PA, pp. 22-27, July 1987.

[6] S. Seshu and D. N. Freeman, ''The Diagnosis of Asynchronous Sequential Switching Systems,'' *IEEE Trans. Electronic Computers*, Vol. EC-11, pp. 459-465, August, 1962.

[7] P. R. Schneider, ''On the Necessity to Examine D-Chains in Diagnostic Test Generation - An Example,'' *IBM J. Res. & Dev.*, Vol. 11, p. 114, January 1967.

[8] *TTL Data Book for Design Engineers, First Edition*, Texas Instruments, Inc., Dallas, Texas, 1973. pages 381-391

[9] ''Special Session: Recent Algorithms for Gate-Level ATPG with Fault Simulation and their Performance Assessment,'' *Proc. 1985 IEEE Int. Symp. Circuits & Systems (ISCAS)*, Kyoto, Japan, pp. 663-698, June 1985, Session Organizers: F. Brglez (Chairman) and H. Fujiwara.

[10] P. Goel, ''An Implicit Enumeration Algorithm to Generate Tests for Combinational Logic Circuits,'' *IEEE Trans. Comp.*, Vol. C-30, pp. 215-222, March 1981.

[11] T. Lin and V. D. Agrawal, ''A Test Generator for Scan-Design VLSI Circuits,'' *Proc. AT&T Conf. Electronic Testing*, Jamesburg, NJ, October 1986.

[12] O. H. Ibarra and S. K. Sahni, ''Polynomially Complete Fault Detection Problems,'' *IEEE Trans. Comp.*, Vol. C-24, pp. 242-249, March 1975.

[13] S. Mallela and S. Wu, ''A Sequential Circuit Test Generation System,'' *Proc. Int. Test Conf.*, Philadelphia, PA, pp. 57-61, November 1985.

[14] H. T. Ma, Private communication.

[15] R. A. Marlett, ''An Effective Test Generation System for Sequential Circuits,'' *Proc. Des. Auto. Conf.*, Las Vegas, Nevada, pp. 250-256, June 1986.

Chapter 7

TEST GENERATION IN CONCURRENT FAULT SIMULATOR

In Chapter 6, we proposed a modified form of simulation to compute the cost functions. This method has proven successful for generating tests for combinational and sequential (synchronous and asynchronous) circuits. However, the threshold-value model used in simulation increases the computation complexity. In this chapter, we will develop another application of the directed-search methodology. We show an effective use of the approach in a concurrent gate-level fault simulator [1,2,3]. Additionally, the approach allows us to develop several new applications for a fault simulator. By defining various cost functions, vectors can be obtained for different purposes like initialization or fault coverage enhancement.

7.1. THREE-PHASE TEST GENERATION

The test generation process can be subdivided into three phases. Fig. 7.1 is the flow chart of the test generation system. In Phase 1 initialization vectors are generated. The purpose of these vectors is to bring flip-flops in the circuit in known states irrespective of the starting state. If designer-supplied initialization vectors are available, or the circuit has hardware features for automatic initialization, then Phase 1 can be skipped.

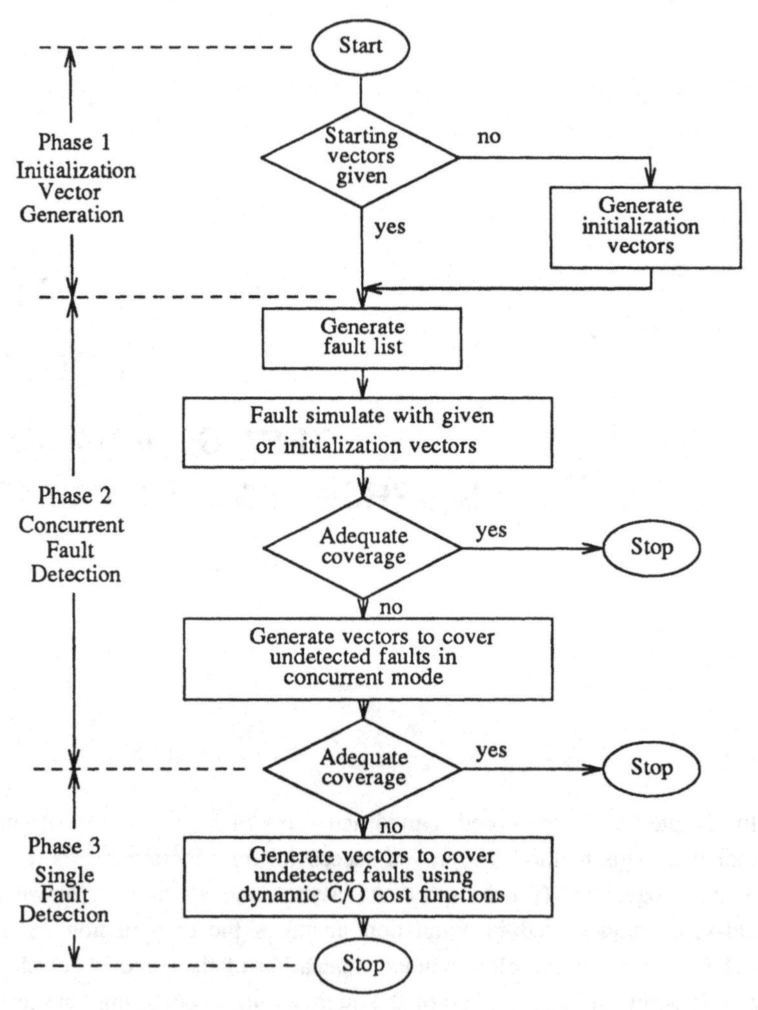

Fig. 7.1 Flow chart of the test generation system.

Phase 2 begins with the vectors that are either supplied by the designer or generated in Phase 1. A fault list is generated in the conventional manner. For example, this list may contain all single stuck faults or a subset of such faults. All faults in this list are simulated using a fault simulator. If the coverage is adequate, the test generation would stop. Otherwise, tests are generated with all undetected faults as targets. In the initial stages of test generation, the fault list is usually long and the objective of this phase is to generate tests effectively by targeting all undetected faults concurrently.

At the end of the Phase 2, if the fault coverage has not reached the required level then Phase 3 is initiated. In this phase, test vectors are generated for a single fault target. Vector generation is guided by dynamically computed controllabilities and observabilities.

7.2. CONCURRENT TEST GENERATION ALGORITHM

In this section, we describe the details of the cost functions and the procedures used in the three phases of test generation developed for a logic simulator.

7.2.1. Phase 1 - Initialization

In this phase, the cost is defined simply as the number of flip-flops that are in the unknown state. Initially, the cost may be equal to the number of flip-flops in the circuit. The goal in the initialization phase is to reduce this cost to 0. Once the circuit is initialized, the test generator may switch to the test generation phase. It is worth noting that the cost function for initialization is derived only from good circuit simulation results and is not related to the faulty circuit behavior. If the circuit is very complicated and hard to initialize, we can relax the criterion for switching to the next phase by allowing a small number of flip-flops, say, 10% uninitialized.

Using this cost function, we drive the circuit into the easiest initialized state instead of any specified state. Details of Phase 1 are shown in Fig. 7.2. All flip-flops are assumed to be in the unknown state at the beginning and the cost function is equal to the number of flip-flops in the circuit. Before applying a trial vector, signal states are saved so that we can return to that state if the trial vector is not accepted. To start the process, any trial vector (randomly generated or user-supplied) can be used. We call this first trial vector as the "current vector". Subsequent trial vectors are generated by changing the bits of the current vector. The integer n is used to denote the bit position in the input vector that is changed. The clock bits (only in the synchronous mode) are treated separately. The user specifies the clock sequence. During simulation, the input data bits are kept fixed whenever the given clock sequence is applied. In combinational or asynchronous sequential circuits, all input bits are treated as data.

After simulation of a trial vector, the "trial cost" is computed as the number of flip-flops that are in the unknown state. If the trial cost is lower than the current cost indicating that the trial vector has initialized some flip-flops that were previously in the unknown state, then the trial vector is saved. If the trial cost is zero, then the initialization phase is complete. Otherwise, the current cost is updated, signal states are saved, the accepted trial vector becomes the current vector, n is set to 1, a new trial vector is generated by changing the nth data bit, and the process of simulation, cost function calculation, etc., is repeated.

A trial vector is not accepted as an initialization vector if the corresponding trial cost is not lower than the current cost. In that case, the bit number n is advanced and the process is repeated with a new trial vector. When all bits of a current vector have been changed without lowering of the cost, this process will stop indicating that an initialization is impossible with this scheme, i.e., a local minimum is reached. We can restart with a new randomly selected current vector and repeat the process of Fig 7.2.

7.2.2. Phase 2 - Concurent Fault Detection

The test generation begins with a circuit that is initialized and several faults are already activated but not detected, i.e., many fault effects are present at internal nodes of the circuit but have not been propagated to primary outputs. We define a cost function for each fault as the shortest distance from any fault effect caused by that fault to any primary output. The distance here is simply the number of logic gates on the path. The smaller the cost, the closer the fault is to being detected. When a fault is detected, its cost will be zero. In test generation, we reduce the cost by propagating the fault effect forward, gate by gate, until finally reaching a primary output. If the fault is not activated, i.e., no fault effect is present anywhere in the circuit, then the cost is defined as infinite.

Figure 7.3 gives a simple example to illustrate how the *distance* cost function works. The given fault is signal A stuck-at-1 and the initial vector is 000. The fault effect appears at signal A, thus, initial cost is 2. After simulating three trial vectors, the search terminates and a test is found.

In general, we may have several undetected faults. We compute the cost C_i for each fault i for some input vector and internal state. We then compute the cost C'_i for a candidate trial vector. Comparing C_i and C'_i, we can determine whether to accept the candidate vector or reject it. It is worth noting that we are dealing with two *lists* of cost functions instead of two numbers. That is, the search for tests is guided by a group of faults instead of a single target fault. We can devise simple rules to determine the acceptance of a vector. For example, if the

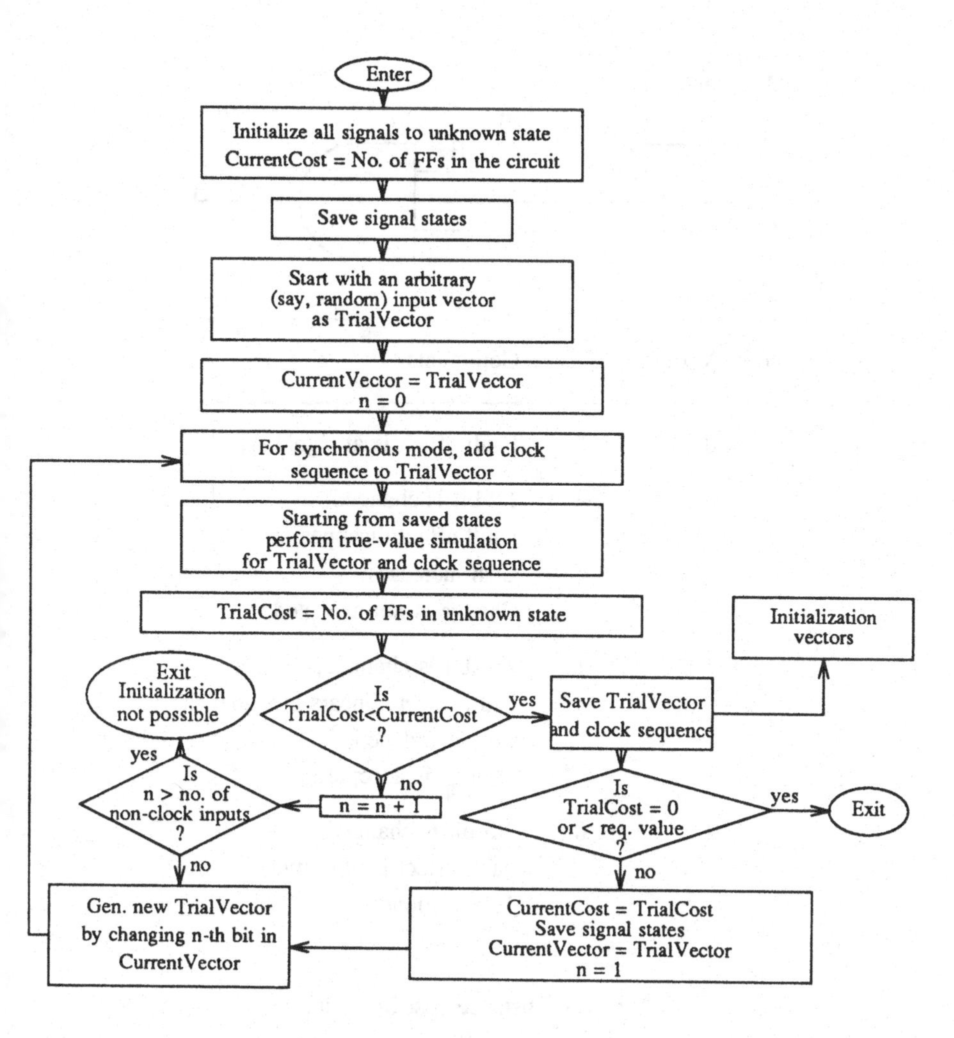

Fig. 7.2 Generation of initialization vectors (Phase 1).

combined cost of 10% of the lowest cost undetected faults is found to decrease, then the new vector may be accepted. Using this cost function, both simulation and test generation can be performed concurrently.

Targeting Multiple Faults. In our experience of running fault simulation, we have found that for most circuits, the test vectors for all stuck-at faults are usually clustered instead of being evenly distributed in the input vector space. Figure 7.4 shows the input vector space with circles representing tests for undetected faults from the fault list. Starting with any initial vector, if the search is directed by the cost of one single fault, it will follow the direction indicated by the dotted line. This search will be rather inefficient and may produce too many test vectors.

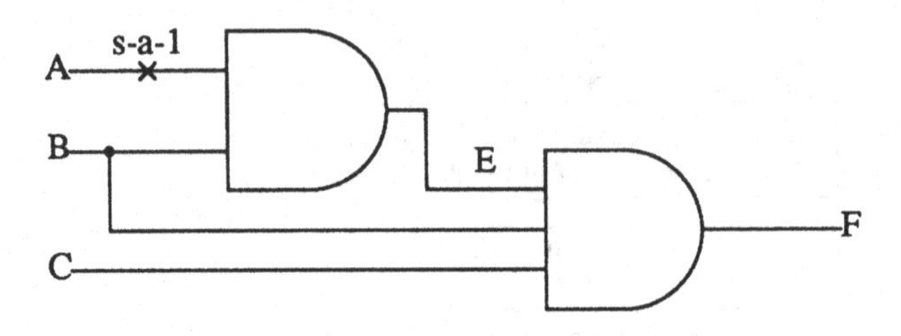

Trial Vector	Cost	Comments
0 0 0	2	Fault effect is at signal A;
1 0 0	∞	1st bit is changed; Fault effect disappears; Cost increases; Change is not accepted;
0 1 0	1	2nd bit is changed; Fault effect appears at signal E; Cost is reduced; Change is accepted;
0 1 1	0	3rd bit is changed; Fault effect is at output; Test is found;

Fig. 7.3 Distance cost function.

On the other hand, if we concurrently target a set of faults, the search will tend to cover large clusters of faults by following the direction indicated by the solid line.

If flip-flops are modeled as functional primitives, we treat them differently from individual gates like AND or OR. Propagating a fault through a gate only needs setting appropriate values at the inputs of the gate. In contrast, propagation through a flip-flop requires first setting the appropriate value at its data input and then activating the clock signal. In cost computation, therefore, a large constant, say 100, is assigned to a flip-flop as its distance contribution.

The cost function used in this phase can not provide any discrimination if the fault is not activated; the cost will be infinity and will be quite useless in the

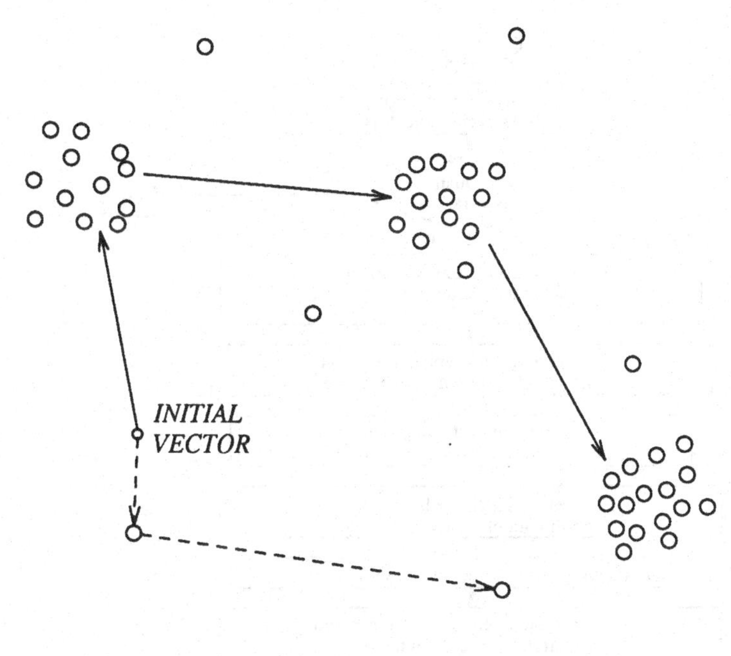

- -> Search directed by a single target fault
—> Search directed concurrently by a set of faults

Fig. 7.4 Multi-target guidence.

search process. Also, if the circuit under test is very shallow, i.e., the number of levels of gates from primary inputs to primary outputs is small, this cost function will be less effective. We will define another cost function for such situations in the next section. Once the test generator finds that the fault coverage can not be improved further by the present cost function, it will automatically switch to the next cost function.

The flow chart of Phase 2 is shown in Fig. 7.5. Generation of trial vectors and minimization is performed in a manner similar to the initialization phase.

The concurrent phase stops when all single bit changes in a current vactor produce no cost reduction. This will normally happen when a small number of faults are left in the target set. The test vectors for these faults sparsely populate the vector space and therefore, the collective cost function does not provide any significant guidance.

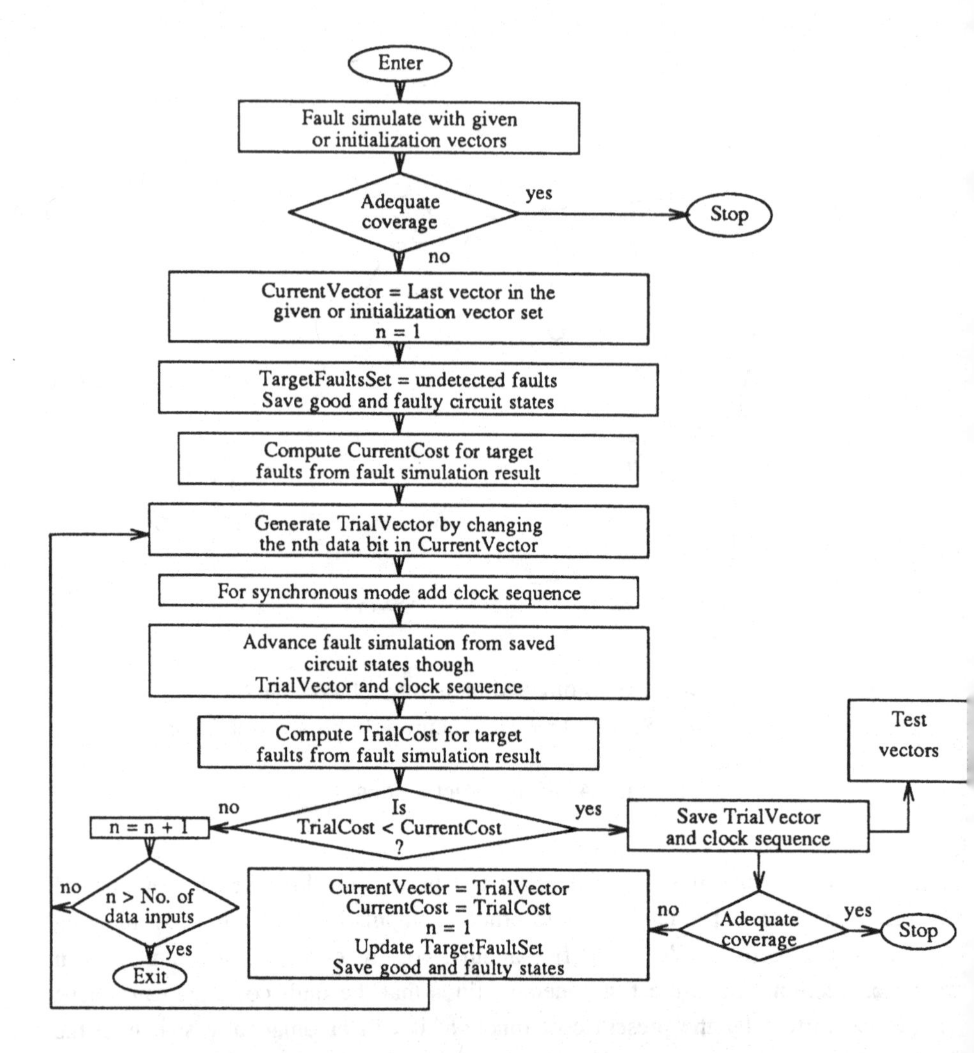

Fig. 7.5 Concurrent fault detection (Phase 2).

7.2.3. Phase 3 - Single Fault Detection

This cost function is based on a SCOAP-like testability measure [4]. There is a difference, however. Instead of computing the measure statically, we dynamically compute it depending on the current state†. Also, the dynamic testability measure is used to compare the suitability of two vectors for detecting a target fault, i.e., it is not used in an absolute sense to assess the testability of the circuit.

† see footnote on page 58.

We define $DC\,1(i)$ and $DC\,0(i)$ as dynamic 1 and 0 controllabilities of node i. These are related to the *minimum* number of primary inputs that must be *changed* and the *minimum* number of additional vectors needed to control the value of node i to 1 or 0. We call the number of inputs to be changed as the dynamic combinational controllability (DCC) and the number of vectors as the dynamic sequential controllability (DSC). These measures are dynamic because they depend upon the current state of the circuit. Since we want to keep the test sequence short, DSC is weighted heavier than DCC. For example, $DC\,1(i)$ and $DC\,0(i)$ could be the weighted sums of DCC and DSC, say, DSC times 100 plus DCC.

7.2.3.1. Dynamic Controllability Measures

If the current logic value on node i is 1, then $DC\,1(i)$ is defined as:

$$DC\,1(i)\ \big|_{V(i)=1} = 0$$

where $V(i)$ is the logic value on node i. Similarly, if the current logic value on node i is zero,

$$DC\,0(i)\ \big|_{V(i)=0} = 0$$

This definition follows from the fact that no input change is needed to justify a 1(0) on node i if the value is already 1(0). Under other conditions, $DC\,1(i)$ and $DC\,0(i)$ will assume non-zero values. For example, for the output line i of an AND gate with m inputs, $DC\,1(i)$ and $DC\,0(i)$ are computed as:

$$DC\,1(i)\ \big|_{V(i)=0\ or\ X} = \sum_{j=1}^{m} DC\,1(k_j)$$

$$DC\,0(i)\ \big|_{V(i)=1\ or\ X} = \min_{1 \le j \le m} DC\,0(k_j)$$

where k_j is the j-th input line of the gate. Here, "min" means the minimum of m quantities and its use is similar to that in SCOAP [4].

As explained above, the controllability of sequential elements is weighted heavier. Dynamic controllabilities for a flip-flop output i is defined as:

$$DC\,1(i)\ \big|_{V(i)=0\ or\ X} = DC\,1(d) + K$$

$$DC\,0(i)\ \big|_{V(i)=1\ or\ X} = DC\,0(d) + K$$

where d is the input data signal of the flip-flop and K is a large constant, say, 100.

7.2.3.2. Activation Cost and Propagation Cost

In order to detect a stuck fault, the test generator must first find a sequence of vectors to activate the fault, i.e., set the appropriate value (opposite of the faulty state) at the fault site and then find another sequence to sensitize a path to propagate the fault effect to a primary output. Thus, the cost function should reflect the effort needed for activating and propagating the fault. The activation cost, $AC(i_j)$, of node i stuck-at-j fault is defined as:

$$AC(i_j) = DCv(i)$$

where v is 0 if node i is stuck-at-1 and v is 1 if node i is stuck-at-0. This follows from the consideration that the cost of activating a stuck-at-0 (stuck-at-1) fault is the cost of setting up a 1 (0) at the fault site.

The propagation cost is basically a dynamic observability measure. For example, for a fanout stem i with n fanout branches, the propagation cost is:

$$PC(i) = \min_{1 \le k \le n} PC(i_k)$$

where i_k is the k-th fanout branch of i. For an input signal i_a of an AND gate whose output signal is i, we have

$$PC(i_a) = PC(i) + \sum_{\substack{1 \le k \le n \\ k \ne a}} DC1(i_k)$$

where n is the number of inputs to the gate. Similar formulas are easily derived for other types of gates.

The cost function for test generation for a single target fault is derived from the activation cost and the propagation cost defined above. For an undetected fault F, line i stuck-at-j, that is not activated, the cost is defined as:

$$Cost(F) \mid_{F \text{ not activated}} = K1 \times AC(i_j) + PC(i)$$

Where $K1$ is a large constant that determines the relative weighting of the two costs. If the fault has been activated, and N_F is the set of the nodes where the fault effect appears, then

$$Cost(F) \mid_{F \text{ activated}} = \min_{i \in N_F} PC(i)$$

Notice that in this cost function reconvergent fanouts are ignored. This approximation provides computational simplicity but may occasionally result in failure to detect a fault.

The flow chart of Phase 3 is shown in Fig. 7.6. In this phase, activation and

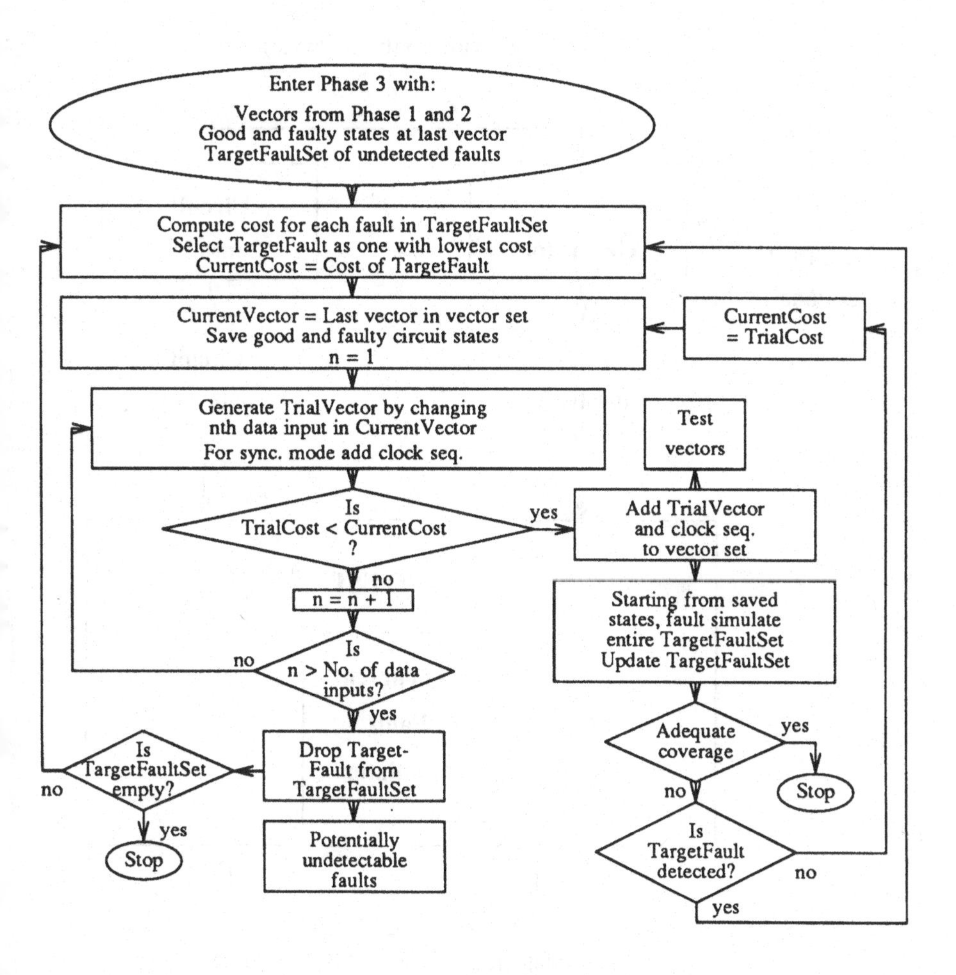

Fig. 7.6 Single fault detection (Phase 3).

propagation costs, as discussed above, are computed for each fault in the list. The lowest cost fault is targeted for detection. New input vectors are created to further lower the cost of the target fault until it is detected or the search is abandoned due to a local cost minimum. As new vectors are added to the sequence, concurrent fault simulation simultaneously eliminates any other detected faults from the list. This phase ends when either an adequate coverage is achieved or all faults that were left undetected at the end of phase 2 have been processed.

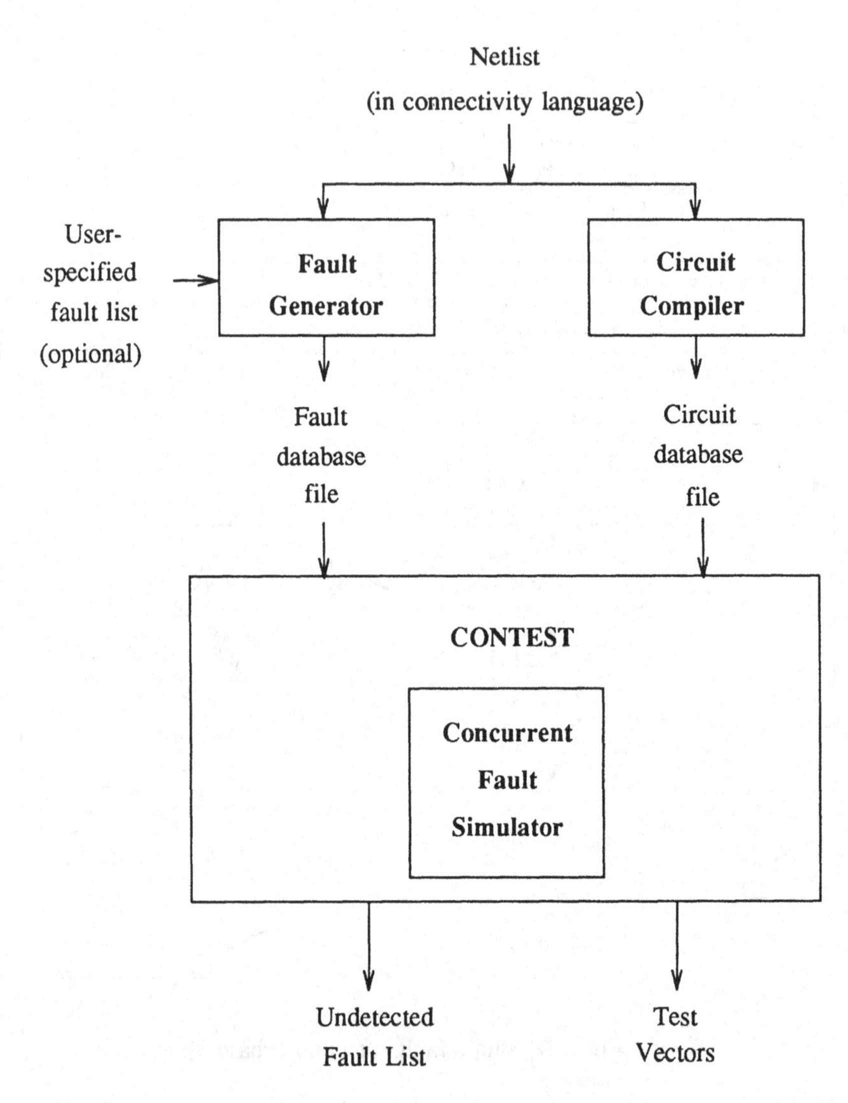

Fig. 7.7 The CONTEST system.

7.3. PROGRAM

Figure 7.7 shows a test generation system that is very similar to the one shown in Fig. 6.8. The inputs, outputs and options of the new system are exactly the same as those discussed in Chapter 6. Circuit compiler and fault generator are also the same. The test generator, CONTEST (CONcurrent TEst generator for Sequential circuit Testing), is built on top of a logic level concurrent fault simulator. It accepts logic-level circuit description in a hardware description language. The test generator also works in two modes: *Synchronous mode* and *Asynchronous mode*. In the synchronous mode, clock signals and their transition sequence within a period must be specified. The test generator follows each change in primary inputs by a clock sequence. In the asynchronous mode, no clock signal is identified and the test generator treats all primary inputs alike. For circuits that are largely synchronous with a limited amount of asynchronous circuitry, test generation in the synchronous mode is recommended. If the coverage by this mode is inadequate then, for the remaining faults, asynchronous mode can be used. This is because the speed of test generation depends upon the number of primary inputs that must be manipulated. In the synchronous mode, clock signals are prespecified and are not manipulated.

A fault list is another input to the test generator. CONTEST contains an event-driven concurrent fault simulator. Race analysis in feedback structures is automatic and is performed through a special modeling feature [5] as discussed in Chapter 6. By default, potentially detectable faults (that produce an unknown faulty output) are considered detected. This option can be turned off by the user. If the number of changes in a signal for the same input vector exceeds a prespecified number, then the simulator assumes oscillation and sets the signal to the unknown state.

In Phase 1, User can specify the acceptable percentage of uninitialized flip-flops. The default is 10 percent. Also, Phase 2, that normally follows Phase 1, can be independently run if the user supplies functional vectors or initialization vectors. Accoring to our experience, Phase 2 can achieve a coverage of 65 to 85 percent. Phase 3 can also be independently run if the size of the given fault list is relatively small.

7.4. EXPERIMENTAL RESULTS

Before discussing test generation for larger circuits, we will illustrate some features of CONTEST by example circuits having typical sequential behavior.

Example 1: The circuit in Fig. 7.8, taken from Marlett's paper [6], was originally used by Miczo [7] to illustrate the problem of initialization. CONTEST generated a four-vector initialization sequence: 00, 01, 00, 01. Starting in the unknown state, this sequence leaves the circuit in the 00 state. To observe how the test generator will move $Q1$ to 1, we generated a test for the fault $Q1$ stuck-at-0. Since it is a single target fault, CONTEST used Phase 3 to generate four more vectors (10,11,00,01) to detect it.

In a separate run, a set of 18 vectors was generated to detect 24 out of 25

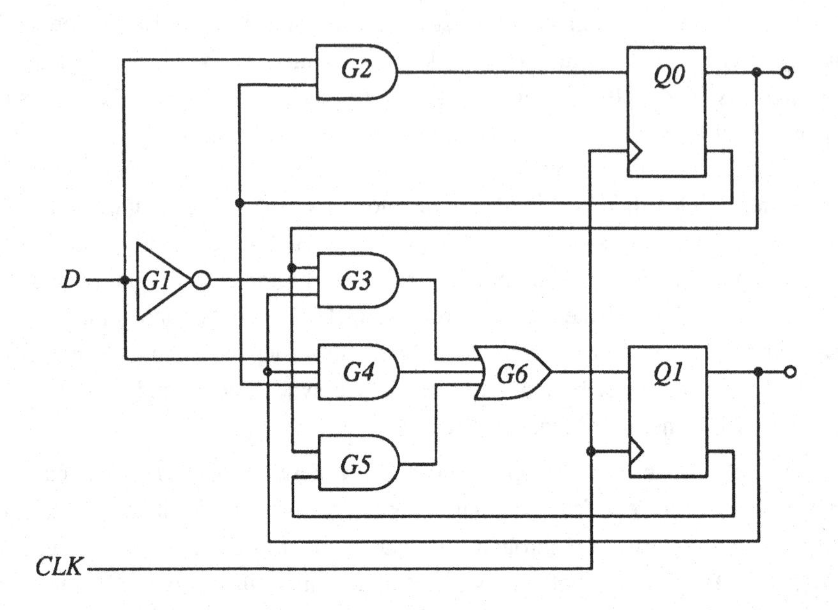

Fig. 7.8 Circuit of Example 1.

faults. The remaining fault, third input (from $Q\,1$) of $G\,3$ stuck-at-1, is redundant.

Example 2: Muth [8] used the asynchronous circuit of Fig. 7.9 to illustrate the necessity of a nine-value model when a combinational test generation method is employed. Following Muth's illustration, we ran CONTEST to generate a test sequence for the fault "*d* stuck-at-1". Four vectors were produced: 000, 100, 101, 111. The last two vectors are the same as given in Muth's paper [8]. Two extra vectors are generated here because CONTEST starts with an arbitrary 000 vector and then brings the circuit to the appropriate state.

Example 3: A test sequence for the four faults at the inputs of the NAND gate in the circuit of Fig. 7.10 was generated by CONTEST. This sequence had fourteen vectors: 00, 01, 10, 11, 00, 01, 10, 11, 00, 01, 10, 11, 00, and 11. The first eight vectors, generated in Phase 1, initialize the circuit to the 0101 state. Since the fault list had only four faults, the program switched to Phase 3. One input stuck-at-1 fault was detected at the 10th vector, both input stuck-at-0 faults were detected at the 12th vector, and the other input stuck-at-1 fault was detected at the 14th vector.

Next, we give CONTEST results for eight sequential circuits. The circuit characteristics are shown in Table 7.1. Six of these circuits are the same as shown in Table 6.5. Of the two new ones, MI is a finite-state machine implemented with

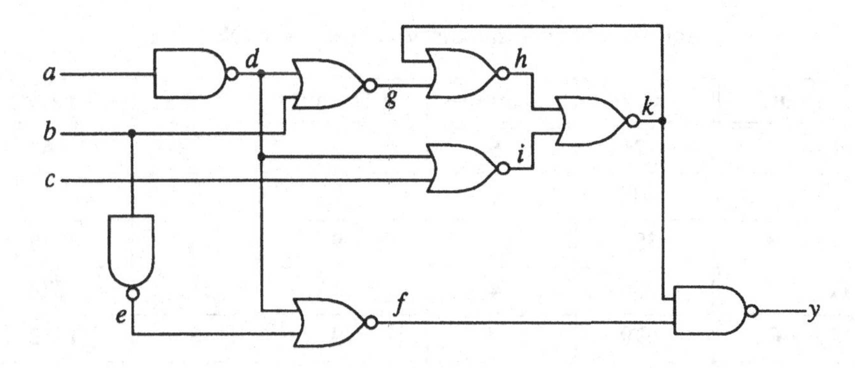

Fig. 7.9 Muth's circuit used in Example 2.

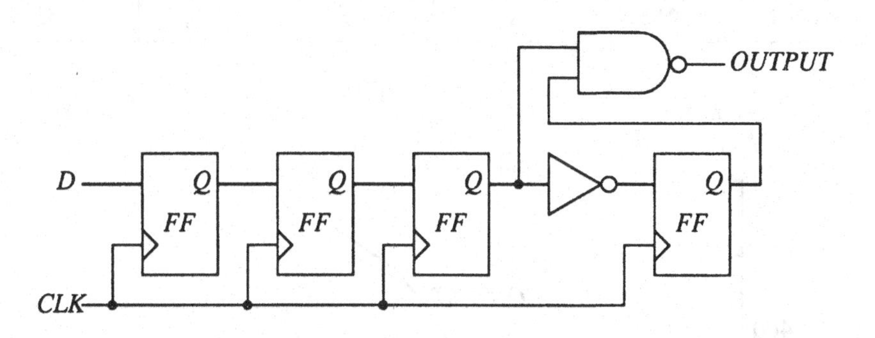

Fig. 7.10. Circuit used in Example 3.

random logic. CHIP-B is a custom chip with standard cell design and is synchronously clocked with the exception of one asynchronous flip-flop.

Fig. 7.11 is a sample of CONTEST results. These vectors were generated for the TLC circuit. First four vectors were generated in the initialization phase. Test generator switched to Phase 3 after generating 550 vectors when the coverage was 85%. The coverage characteristic of random vectors, shown in the figure for comparison, is typical for sequential circuits. Fig. 7.12 shows the coverage of vectors generated by CONTEST, TVSET and a random vector generator for the MULT4 circuit. MULT4 is a mildly sequential circuit, therefore, the coverage of random vectors is much higher than that for TLC.

Table 7.2 shows the results obtained by CONTEST for the eight circuits of Table 7.1. For comparison, results of TVSET and STG [9], are also included from Table 6.6. In general, CONTEST and TVSET have similar performance which is are better than STG. Fault coverages reported in Table 7.2 were verified through

Table 7.1 Sequential Circuits used for CONTEST

Circuit	# Gate	# Input	# Output	# FFs	# Faults
MANNY	26	5	3	7‡	67
SSE	207	8	7	6	454
MULT4	382	10	9	15	540
TLC	355	4	6	21	772
PLANET	690	9	19	6	1582
MI	779	15	14	18	1629
CHIP-A	1112	13	13	39	1643
CHIP-B	1539	17	5	73§	2533

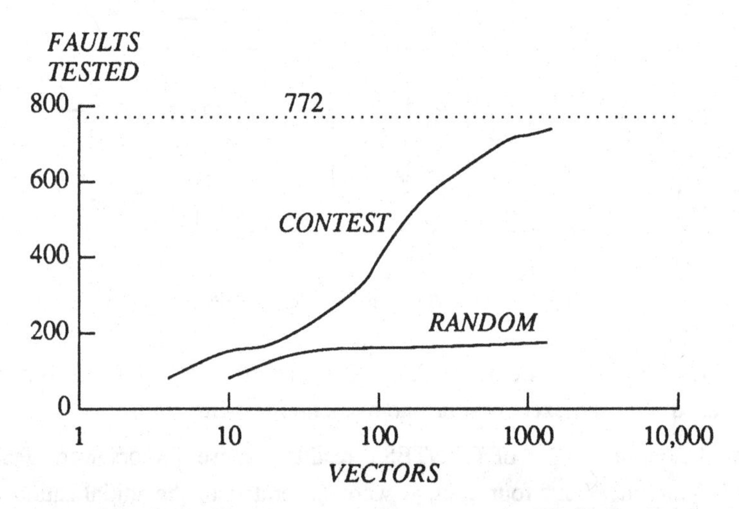

Fig. 7.11 Fault coverage for vectors for TLC circuit.

an MOS fault simulator [10].

Run time estimates of CONTEST and TVSET are conservative as they are based on our experimental implementation. Their speed, that largely depends on the speed of the fault simulator, can be significantly improved.

Table 7.3 summarizes the result of test generation for CHIP-A obtained by

‡ Asynchronous flip-flops formed by cross-coupled NAND gates.

§ Includes one asynchronously controlled flip-flop.

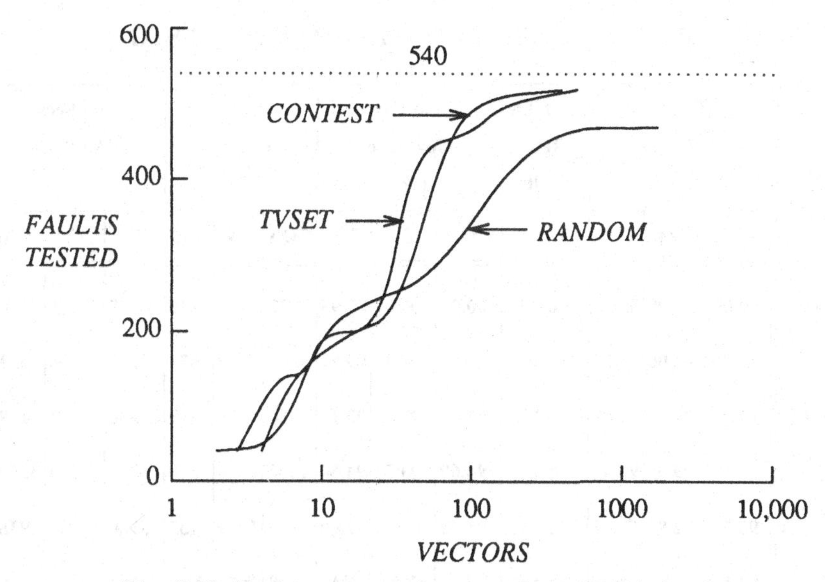

Fig. 7.12 Fault coverage of vectors for MULT4 circuit.

four different test generators: CONTEST, TVSET, STG, and a commercially available test generation program [6] marketed by the HHB Systems (see also Sec. 6.4.2). Fault coverages reported were verified through an MOS fault simulator.

Since both STG and the commercial program use an extended backtrace technique for an efficient implementation of the path sensitization procedure, the number of vectors they generated is about the same. CONTEST and TVSET, on the other hand, consistently generated more vectors than the other two programs. This is a consequence of the one-bit change heuristic. In general, neighboring vectors in a CONTEST-generated or TVSET-generated sequence have fewer bit changes than in sequences generated by the other two methods. As a result, more vectors are needed to take the circuit to a desired state starting from some given state. In practical testing environment, this may be an advantage because too many simultaneous input changes can produce hazards in the logic or produce power supply fluctuation due to current surge. The single-bit change strategy has also been used by other workers [11, 12].

It is our belief that the run time of CONTEST and TVSET programs can be significantly improved since it totally depends upon the fault simulation program used. The present versions use experimental fault simulators.

The fault coverages obtained by CONTEST are slightly better than those obtained by TVSET. This is a consequence of the concurrent fault detection

Table 7.2 Results of CONTEST, TVSET and STG

Circuit	Fault Coverage (%)			Provable Red. Faults	Total Fault Coverage (%)			# Test Vectors			CPU Secs. (VAX 8650)		
	CON*	TV†	STG	(%)	CON	TV	STG	CON	TV	STG	CON	TV	STG
MANNY	100.00	100.00	83.95	0	100.00	100.00	83.95	32	35	219	5	5	NA
SSE	83.26	81.06	83.20	16.30	99.56	97.36	99.50	561	590	676	291	398	1134
MULT4	97.04	96.67	92.78	0.37	97.41	97.04	93.15	364	388	148	838	1068	1490
TLC	94.69	93.52	94.64	NA	94.69	93.52	94.64	1256	1256	5340	3312	4514	32590
PLANET	95.13	92.92	57.71	3.29	98.42	96.21	61.00	1439	1469	132	3120	4621	19388
MI	94.53	NA	NA	NA	94.53	NA	NA	1358	NA	NA	1261	NA	NA
CHIP-A	93.73	93.06	84.11	4.37	98.10	97.43	88.48	1031	924	384	98432	89351	NA
CHIP-B	91.28	NA	NA	NA	91.28	NA	NA	1034	NA	NA	77904	NA	NA

Table 7.3 Results of CHIP-A

Circuit	Total Fault Coverage (%)				# Test Vectors				CPU Secs. (VAX 8650)			
	CON	TV	STG	HHB	CON	TV	STG	HHB	CON	TV	STG	HHB
CHIP-A	98.10	97.43	88.48	95.12	1031	924	384	330	98432	89351	NA	NA

strategy. It is found that most of the faults that are covered by CONTEST but not by TVSET are detected in Phase 2 (concurrent fault detection phase) of CONTEST. In general, the run time of TVSET is slightly longer than that of CONTEST. There are two reasons for this: (1) binary values are used in CONTEST

* CONTEST
† TVSET

while integers are required in TVSET; and (2) there are more fault events processed in a threshold-value simulator than in a logic fault simulator.

In view of the results of TVSET and CONTEST presented here, our unit Hamming distance heuristic seem to work well. In most cases, we were *unable* to manually generate tests for the faults that were left undetected by CONTEST and TVSET. Finding redundancies in sequential circuits was even harder. When the final fault coverage is lower than the desired goal we see two options. The first option is to start with a different (randomly selected) vector and attempt generation of tests for the undetected faults. The second option is to expand the one-bit change heuristic to include two-bit, three-bit, . . . changes. One should, however, expect a rapid increase in the amount of computations.

7.4.1. Empirical Complexity Analysis of CONTEST

Even though the problem of guaranteed test generation is NP-complete, it will be useful to analyze the complexity of the algorithm and heuristics by using experimental results. In the new method, for synchronous sequential circuits, all primary inputs must be flipped once to generate one test vector in the search process. Assuming that the simulation complexity per vector is proportional to the number of gates, the complexity of generating one test vector is proportional to $M \times G$ where M is the number of primary inputs and G is the number of gates. For a given circuit, let L be the average length of test sequence required to detect a fault. The complexity of generating this test sequence is proportional to $L \times M \times G$.

For the same circuit, let L' be the average test sequence length generated for detecting a fault by the conventional iterative-array approach. Since L' copies of the combinational portion are made in this approach, the size of the corresponding combinational model is $L' \times G$. For a combinational circuit of average complexity, D-Algorithm or PODEM run as $O(g^n)$ where g is the number of gates in the circuit and n is empirically found to be between 1.8 and 2. Thus, the average complexity of the iterative-array approach for generating a test sequence to detect a fault is proportional to

$$(L' \times G)^n .$$

From the experimental results, L and L' can be computed by dividing the total number of generated test vectors by the number of faults that were used as targets during test generation. The number of target faults is usually much smaller than the total number of faults in the circuit because many faults are covered by fault simulation. Assuming L and L' to be are nearly identical and taking n as 2, the ratio of the complexity of the conventional approach to the directed-search is

$$L \times \frac{G}{M}.$$

This expression clearly shows the problem with the conventional method when long test sequences are to be generated. Further, as the circuit size increases, G increases more rapidly than M. Thus, the directed-search will have better potential to handle larger circuits.

We use the TLC Circuit to illustrate the run time advantage of the directed-search method. Even though there are only 21 flip-flops in this circuit, it is a highly sequential circuit with an internal counter and the ratio of logic to primary inputs is high (only 4 primary inputs). For some faults, more than 200 vectors are needed to initialize the circuit to appropiate states needed to activate and propagate the fault effects to primary outputs. That means more than 200 copies of the combinational portion are created in the iterative-array model. This is one of the reasons why STG required much more CPU time than CONTEST and TVSET for the TLC circuit.

REFERENCES

[1] K. T. Cheng, V. D. Agrawal, and P. Agrawal, "Use of a Concurrent Fault Simulator for Test Vector Generation," *Proc. AT&T Conference on Electronic Testing*, Princeton, NJ, pp. 23-28, October 1987.

[2] V. D. Agrawal, K. T. Cheng, and P. Agrawal, "CONTEST: A Concurrent Test Generator for Sequential Circuits," *25th Design Automation Conf.*, Anaheim, CA, pp. 84-89, June 1988.

[3] V. D. Agrawal, K. T. Cheng, and P. Agrawal, "A Directed Search Method for Test Generation Using a Concurrent Simulator," *IEEE Trans. on CAD*, Vol. 8, pp. 131-138, February 1989.

[4] L. H. Goldstein, "Controllability/Observability Analysis of Digital Circuits," *IEEE Trans. on Circuits and Systems*, Vol. CAS-26, pp. 685-693, Sept 1979.

[5] E. G. Ulrich and T. Baker, "Concurrent Simulation of Nearly Identical Digital Networks," *Computer*, Vol. 7, pp. 39-44, April 1974.

[6] R. A. Marlett, "An Effective Test Generation System for Sequential Circuits," *Proc. Des. Auto. Conf.*, Las Vegas, Nevada, pp. 250-256, June 1986.

[7] A. Miczo, "The Sequential ATPG: A Theoretical Limit," *Proc. Int. Test*

Conf., Philadelphia, PA, pp. 143-147, October 1983.

[8] P. Muth, "A Nine-Valued Circuit Model for Test Generation," *IEEE Trans. Comp.*, Vol. C-25, pp. 630-636, June 1976.

[9] S. Mallela and S. Wu, "A Sequential Circuit Test Generation System," *Proc. Int. Test Conf.*, Philadelphia, PA, pp. 57-61, November 1985.

[10] C. Y. Lo, H. N. Nham, and A. K. Bose, "Algorithms for an Advanced Fault Simulation System in MOTIS," *IEEE Trans. Computer Aided Design*, Vol. CAD-6, pp. 232-240, March 1987.

[11] S. Seshu and D. N. Freeman, "The Diagnosis of Asynchronous Sequential Switching Systems," *IEEE Trans. Electronic Computers*, Vol. EC-11, pp. 459-465, August, 1962.

[12] J. Rajski and H. Cox, "A Method of Test Generation and Fault Diagnosis in Very Large Combinational Circuits," *Proc. Int Test Conf.*, Washington, D.C., pp. 932-943, September 1987.

Chapter 8

CONCLUSIONS AND FUTURE WORK

Test vector generation for general sequential circuits is considered to be the hardest problem in testing. In order to simplify testing, design for testability (DFT) is often recommended. However, because of the high, and often unacceptable, overhead of the DFT methods like scan design and BIST, a reliable test generator for sequential logic is really desirable. Most of the previous test generation approaches have two problems: (1) they rely on backtracking, and (2) they ignore circuit delays. For sequential circuits, backtracking is required over several time frame expansions and could easily become unmanageable. Neglecting delays can make a test impossible or produce a test that causes race in the circuit. The importance of the sequential test generation problem and the deficiencies of previous approaches motivated this research.

We have formulated the principle of simulation-based directed-search for test generation. Two practical and completely automatic test generation algorithms for sequential and combinational circuits are presented using the new principle. These algorithms are implemented in the programs CONTEST and TVSET. The use of an event-driven fault simulator adds to the effectiveness and accuracy. The success of the new approach is a combined effect of many features that further highlight contributions of this work: (1) The use of simulation allows proper consideration of circuit delays, feedbacks, and reconvergent fanouts. Thus, the timing problem in test generation is solved and the vectors generated are guaranteed to be race-free and hazard-free. (2) No explicit backtracking is involved. All decisions during the search are made on the basis of inputs and outputs and no explicit consideration of the internal structure (reconvergent fanouts) is required. Also, no time frame expansion is needed for sequential logic. (3) Asynchronous sequential circuits can be handled since they pose no problem for the fault simulator. (4) Tests can also be generated for any other fault models that can be simulated. (5) CONTEST can be easily implemented in many currently used CAD environments. This is because a concurrent fault simulator is a very popular CAD tool in the industry and the test generator can be incorporated in an existing fault simulator without significant changes to the data structure. (6) The threshold-value simulation combines logic simulation and dynamic (pattern-dependent) testability evaluation and, thus, also provides a unified framework for simulation and test generation. (7) According to the empirical complexity analysis and experimental results, the new approach has the potential to handle VLSI chips.

It is hoped that this work will lead to further research. The future work can be classified into two categories: enhancements to TVSET and CONTEST by adding to their capabilities and improving their performance; and new techniques for cost computation and extension of the threshold-value model to switch-level and function-level circuit representations. It is also desirable to extend the CONTEST algorithm to handle high-level functional blocks. The key problems are: defining the *distances* between inputs and outputs of functional blocks and computing dynamic controllabilities and observabilities through these blocks.

In connection with enhancements to improve performance, several things can be suggested: (1) It is possible to preprocess the circuit to identify those primary inputs which do not influence the detection of the fault under consideration. During the search for a test for that fault such primary inputs could then remain unchanged. (2) One can develop a more sophisticated strategy for generating trial vectors during search. Currently, the input signals are modified one at a time. This strategy is simple and works in most cases. However, a more sophisticated strategy may have a better potential for avoiding local minima. One may still apply the one-bit change heuristic in the first pass and use a more sophisticated mechanism in subsequent passes. In this way, a certain percentage of fault coverage can be obtained within a reasonable computing time and a higher fault coverage can still be reached if longer computing time is affordable. (3) Experimental results show that the undetected faults of TVSET and CONTEST generally form a subset of the undetected faults of STG. Further characterization of such faults and investigation to find the specific circuit structures that make detection difficult is recommended. (4) Implementation of the CONTEST algorithm on a hardware accelerator is also recommended. The test generator can be incorporated into any commercial hardware-based fault simulator. (5) The cost function formulation of the test generation problem hints at a new application of simulated annealing. Even though simulated annealing might be more expensive in terms of computing time, it may still be a good solution to avoid local minima.

Chapter 9

BIBLIOGRAPHY

In preparing this bibliography of material on test generation we have reviewed most major books, journals and conference proceedings published during 1980-88. In addition, many key items published prior to 1980 are also included.

[1] J. Abadir and Y. Deswarte, "Run-Time Program for Self-Checking Single Board Computer," *Int. Test Conf. Digest of Papers*, Philadelphia, PA, November 1982, pp. 205-213.

[2] M. S. Abadir and H. K. Reghbati, "LSI Testing Techniques,"*IEEE Micro*, Vol. 3, pp. 34-50, Feb. 1983.

[3] M. S. Abadir and H. K. Reghbati, "Functional Testing of Semiconductor Random Access Memories,"*ACM Computing Surveys*, Vol. 15, pp. 175-198, September 1983.

[4] M. S. Abadir and H. K. Reghbati, "Functional Test Generation for LSI Circuits Described by Binary Decision Diagrams," *Proc. Int. Test Conf.*, Philadelphia, PA, November 1985, pp. 483-492.

[5] M. S. Abadir and H. K. Reghbati, "Functional Test Generation for Digital Circuits Described Using Binary Decision Diagrams,"*IEEE Trans. Computers*, Vol. C-35, pp. 375-379, April 1986.

[6] M. S. Abadir and H. K. Reghbati, "Functional Test Generation Using Binary Decision Diagrams,"*Comput. Math. Applic.*, Vol. 13, No. 5/6, pp. 413-430, 1987.

[7] Z. Abazi and P. Thevenod-Fosse, "Markov Models for the Random Testing Analysis of Cards,"*Fault-Tolerant Comp. Symp. (FTCS-16) Digest of Papers*, pp. 272-277, July 1986.

[8] S. Aborhey, "Binary Decision Tree Test Functions,"*IEEE Trans. Computers*, Vol. 37, pp. 1461-1465, November 1988.

[9] E. M. Aboulhamid, "Efficient Testing and Truth Table Verification of Unilateral Combinational Iterative Arrays," *Proc. Int. Conf. on CAD (ICCAD)*, Santa Clara, CA, November 1985, pp. 68-70.

[10] M. Abramovici and M. A. Breuer, "Multiple Fault Diagnosis in Combinational Circuits Based on an Effect-Cause Analysis,"*IEEE Trans. Computers*, Vol. C-29, pp. 451-460, June 1980.

[11] M. Abramovici and M. A. Breuer, "Fault Diagnosis in Synchronous Sequential Circuits Based on an Effect-Cause Analysis,"*IEEE Trans. Computers*, Vol. C-31, pp. 1165-1172, December 1982. Also *Fault-Tolerant Comp. Symp. (FTCS-10) Digest of Papers*, Kyoto, Japan, October 1980, pp. 313-315.

[12] M. Abramovici, J. J. Kulikowski, P. R. Menon, and D. T. Miller, "SMART and FAST: Test Generation for VLSI Scan-Design Circuits,"*IEEE Design & Test of Computers*, Vol. 3, pp. 43-54, Aug., 1986. Also *Proc. Int. Test Conf.*, Philadelphia, PA, Nov., 1985, pp. 45-56.

[13] M. Abramovici and P. R. Menon, "A Practical Approach to Fault Simulation and Test Generation for Bridging Faults,"*IEEE Trans. Computers*, Vol. C-34, pp. 658-663, July 1985. Also *Proc. Int. Test Conf.*, Philadelphia, PA, October 1983, pp. 138-142.

[14] C. F. Acken, "Automatic Test Generating Using a Matrix Model of Digital Systems," *Proc. Int. Conf. Circ. Computers (ICCC'82)*, New York, NY, September 1982, pp. 464-467.

[15] M. Adiletta, E. Cooper, and K. Gutfreund, "Automatic Test Generation for Generic Scan Designs," *Proc. Int. Test Conf.*, Philadelphia, PA, November 1985, pp. 40-44.

[16] V. K. Agarwal, "Multiple Fault Detection in Programmable Logic Arrays,"*IEEE Trans. Computers*, Vol. C-29, pp. 518-522, June 1980.

[17] P. Agrawal, "Test Generation at Switch Level," *Int. Conf. Computer-*

Aided Design (ICCAD-84), Santa Clara, CA, Nov. 1984, pp. 128-130.

[18] P. Agrawal and V. D. Agrawal, "Probabilistic Analysis of Random Test Generation Method of Irredundant Combinational Logic Networks,"*IEEE Transaction on Computers*, Vol. C-24, pp. 691-695, July, 1975.

[19] P. Agrawal and V. D. Agrawal, "On Monte Carlo Testing of Logic Tree Networks,"*IEEE Trans. Computers*, Vol. C-15, pp. 664-667, June 1976.

[20] P. Agrawal and S. M. Reddy, "Test Generation at MOS Level," *Int. Conf. on Computer, Systems & Signal Processing*, Bangalore, India, Dec. 1984, pp. 1116-1119.

[21] V. D. Agrawal, "When to Use Random Testing,"*IEEE Trans. Computers*, Vol. C-27, pp. 1054-1055, Nov. 1978. Also see comments by P. B. Schneck and the author's reply in *IEEE Trans. Comp.*, Vol. C-28, pp. 580-581, August 1979.

[22] V. D. Agrawal, "Information Theory in Digital Testing – A New Approach to Functional Test Pattern Generation," *Proc. Int. Conf. Circ. Computers (ICCC-80)*, Port Chester, NY, October 1980, pp. 928-931.

[23] V. D. Agrawal, "An Information Theoretic Approach to Digital Fault Testing,"*IEEE Transaction on Computers*, Vol. C-30, pp. 582-587, August 1981.

[24] V. D. Agrawal, "Statistical Testing," in *Testing and Diagnosis of VLSI and ULSI*, M. Sami and F. Lombardi, Eds., Martinus Nijhof Publishers, Dordrecht, the Netherlands, 1988.

[25] V. D. Agrawal and P. Agrawal, "An Automatic Test Generation System for Illiac IV Logic Boards,"*IEEE Trans. Computers*, Vol. C-21, pp. 1015-1017, September 1972.

[26] V. D. Agrawal and K. T. Cheng, "Threshold-Value Simulation and Test Generation," in *Testing and Diagnosis of VLSI and ULSI*, M. Sami and F. Lombardi, Eds., Martinus Nijhof Publishers, Dordrecht, the Netherlands, 1988.

[27] V. D. Agrawal, K. T. Cheng, and P. Agrawal, "CONTEST: A Concurrent Test Generator for Sequential Circuits," *Proc. 25th Des. Auto. Conf.*, Anaheim, CA, June 1988, pp. 84-89.

[28] V. D. Agrawal, K. T. Cheng, and P. Agrawal, "A Directed Search Method for Test Generation Using a Concurrent Fault Simulator,"*IEEE Trans. on CAD*, Vol. 8, pp. 131-138, February 1989.

[29] V. D. Agrawal, H. Farhat, and S. Seth, "Test Generation by Fault

Sampling,'' *Proc. IEEE Int. Conf. Comp. Design (ICCD-88)*, Rye Brook, NY, October 1988, pp. 58-61.

[30] V. D. Agrawal and D. D. Johnson, ''Logic Modeling of PLA Faults,'' *Proc. Int. Conf. on Computer Design (ICCD-86)*, Port Chester, NY, October 1986, pp. 86-88.

[31] V. D. Agrawal and S. M. Reddy, ''Fault Modeling and Test Generation,'' in *VLSI Handbook*, J. DiGiacomo, Ed., McGraw-Hill, New York, 1989.

[32] V. D. Agrawal and S. C. Seth, *Test Generation for VLSI Chips*, Computer Society Press, Washington, D.C., 1988.

[33] V. D. Agrawal, S. C. Seth, and C. C. Chuang, ''Probabilistically Guided Test Generation,'' *Proc. of Int. Symp. Circuits and Systems (ISCAS)*, May 1985, pp. 687-689.

[34] A. N. Airapetian and J. F. McDonald, ''Improved Test Set Generation Algorithm for Combinational Circuit Control,''*Fault-Tolerant Computing Symp. (FTCS-9) Digest of Papers*, pp. 133-136, Jun. 1979.

[35] S. B. Akers, C. Joseph, and B. Krishnamurthy, ''On the Role of Independent Fault Sets in the Generation of Minimal Test Sets,'' *Proc. Int. Test Conf.*, Washington, DC, September 1987, pp. 1100-1107.

[36] G. Alfs, R. W. Hartenstein, and A. Wodtko, ''The KARL/KARATE System: Automatic Test Pattern Generation Based on RT Level Descriptions,'' *Proc. Int. Test Conf.*, Washington, DC, September 1988, pp. 230-235.

[37] M. A. Annaratone and M. G. Sami, ''An Approach to Functional Testing of Microprocessors,''*Fault-Tolerant Comp. Symp. (FTCS-12) Digest of Papers*, pp. 158-164, June 1982.

[38] D. B. Armstrong, ''On Finding a Nearly Minimal Set of Fault Detection Tests for Combinational Logic Nets,''*IEEE Trans. Electronic Computers*, pp. 66-74, October 1965.

[39] D. S. Barclay and J. R. Armstrong, ''A Heuristic Chip-Level Test Generation Algorithm,'' *Proc. 23rd Des. Auto. Conf.*, Las Vegas, Nevada, June 1986, pp. 257-262.

[40] L. Basto, P. Harrod, and W. Bruce, ''Testing the MC68030 Caches,'' *Proc. Int. Test Conf.*, Washington, DC, September 1987, pp. 826-833.

[41] C. Bellon, A. Liothin, S. Sadier, G. Saucier, R. Velazco, F. Grillot, and M. Issenman, ''Automatic Generation of microprocessor Test Programs,'' *Proc. 19th Des. Auto. Conf.*, Las Vegas, Nevada, June 1982, pp. 566-573.

[42] C. Bellon, C. Robach, and G. Saucier, ''An Intelligent Assistant for Test Program Generation: The Supercat System,'' *Proc. Int. Conf. CAD (ICCAD-83)*, Santa Clara, CA, September 1983, pp. 32-33.

[43] C. Bellon, C. Robach, and G. Saucier, ''VLSI Test Program Generation: A System for Intelligent Assistance,'' *Proc. Int. Conf. Comp. Des. (ICCD'83)*, Port Chester, NY, October 1983, pp. 49-52.

[44] C. Bellon and R. Velazco, ''Hardware and Software Tools for Microprocessor Functional Test,'' *Proc. Int. Test Conf.*, Philadelphia, PA, October 1984, pp. 804-810.

[45] C. Bellon, R. Velazco, and H. Ziade, ''Analysis of Experimental Results on Functional Testing and Diagnosis of Complex Circuits,'' *Proc. Int. Test Conf.*, Washington, DC, September 1988, pp. 64-72.

[46] M. J. Bending, ''Hitest: A Knowledge-Based Test Generation System,''*IEEE Design & Test of Computers*, Vol. 1, pp. 83-92, May 1984.

[47] C. Benmehrez and J. F. McDonald, ''The Subscripted D-Algorithm – ATPG With Multiple Independent Control Paths,'' *ATPG Workshop Proceedings*, 1983, pp. 71-80.

[48] C. Benmehrez and J. F. McDonald, ''Measured Performance of a Programmed Implementation of the Subscripted D-Algorithm,'' *Proc. 20th Des. Auto. Conf.*, Miami Beach, FL, June 1983, pp. 308-315.

[49] N. Benowitz, D. F. Calhoun, G. E. Anderson, J. E. Bauer, and C. T. Joeckel, ''An Advanced Fault Isolation System for Digital Logic,''*IEEE Trans. Comp.*, Vol. C-24, pp. 489-497, May 1975.

[50] R. Beresford, ''Gate-array and Standard-cell Design Methods, Part 3: Test Generation,''*VLSI Design*, Vol. V, p. 46, July 1984.

[51] W. E. D. Beste, ''Using a Software Emulator to Generate and Edit VLSI Test Patterns,''*Electronics Test*, pp. 42-52, March 1984.

[52] D. Bhattacharya and J. P. Hayes, ''High-Level Test Generation Using Bus Faults,''*Fault-Tolerant Comp. Symp. (FTCS-15) Digest of Papers*, pp. 65-70, June 1985.

[53] D. Borrione, P. Camurati, J. L. Paillet, and P. Prinetto, ''A Functional Approach to Formal Hardware Verification: The MTI Experience,'' *Proc. IEEE Int. Conf. Comp. Design (ICCD-88)*, Rye Brook, NY, October 1988, pp. 592-595.

[54] P. Bose, ''DEPLOMAT: A Design Expert for PLA Optimization,

Maintenance and Test,'' *Proc. Int. Conf. Comp. Des. (ICCD-87)*, Rye Brook, NY, October 1987, pp. 292-296.

[55] P. Bose and J. A. Abraham, ''Test Generation for Programmable Logic Arrays,'' *Proc. ACM/ IEEE Design Automation Conf.*, Las Vegas, Nevada, June 1982, pp. 574-580.

[56] D. Brahme and J. A. Abraham, ''Functional Testing of Microprocessors,''*IEEE Trans. Comp.*, Vol. C-33, pp. 475-485, June 1984.

[57] M. Brashler, D. Coleman, and R. Dubois, ''An Integrated IC Test Development System,'' *Proc. Custom Integrated Circuits Conf.*, Rochester, NY, May 1984, pp. 169-171.

[58] R. D. Braun and D. T. Givone, ''A Generalized Algorithm for Constructing Checking Sequences,''*IEEE Trans. Computers*, Vol. C-30, pp. 141-144, February 1981.

[59] M. A. Breuer and M. Abramovici, ''Fault Diagnosis Based on Effect-Cause Analysis,'' *Proc. 17th Des. Auto. Conf.*, Minneapolis, MN, June 1980, pp. 69-76.

[60] M. A. Breuer and A. D. Friedman, *Diagnosis & Reliable Design of Digital Systems*, Computer Science Press, Rockville, MD, 1976.

[61] M. A. Breuer and R. L. Harrison, ''Procedures for Eliminating Static and Dynamic Hazards in Test Generation,''*IEEE Trans. Computers*, Vol. C-23, pp. 1069-1092, October 1974.

[62] R. E. Bryant, ''Graph-Based Algorithms for Boolean Function Manipulation,''*IEEE Trans. Comp.*, Vol. C-35, August 1986.

[63] J. P. Caisso and B. Courtois, ''Fault Simulation and Test Pattern Generation at the Multiple-Valued Switch Level,'' *Proc. Int. Test Conf.*, Washington, DC, September 1988, pp. 94-101.

[64] J. L. Carter, S. F. Dennis, V. S. Iyengar, and B. K. Rosen, ''ATPG via Random Pattern Simulation,'' *Proc. Int. Symp. Circuits and Systems (ISCAS)*, Tokyo, Japan, May 1985, pp. 683-686.

[65] E. Cerny, D. Mange, and E. Sanchez, ''Synthesis of Minimal Binary Decision Trees,''*IEEE Trans. Computers*, Vol. C-28, pp. 472-482, 1979.

[66] S. T. Chakradhar, M. L. Bushnell, and V. D. Agrawal, ''Automatic Test Generation Using Neural Networks,'' *Proc. Int. Conf. CAD (ICCAD-88)*, Santa Clara, CA, November 1988, pp. 416-419.

[67] S. J. Chandra and J. H. Patel, ''A Hierarchical Approach to Test Vector

Generation," *Proc. 24th Des. Auto. Conf.*, Miami Beach, FL, June 1987, pp. 495-501.

[68] S. J. Chandra and J. H. Patel, "Test Generation in a Parallel Processing Environment," *Proc. IEEE Int. Conf. Comp. Design (ICCD-88)*, Rye Brook, NY, October 1988, pp. 11-14.

[69] S. J. Chandra and J. H. Patel, "Experimental Evaluation of Testability Measures for Test Generation,"*IEEE Trans. on CAD*, Vol. 8, pp. 93-97, January 1989.

[70] R. Chandramouli, "On Testing Stuck-Open Faults,"*13th Int. Symp. Fault Tolerant Computing (FTCS-13) Digest of Papers*, pp. 258-265, June 1983.

[71] H. P. Chang, W. A. Rogers, and J. A. Abraham, "Structured Functional Level Test Generation Using Binary Decision Diagrams," *Proc. Int. Test Conf.*, Washington, DC, September 1986, pp. 97-104.

[72] H. Y. Chang, E. G. Manning, and G. Metze, *Fault Diagnosis of Digital Systems*, Wiley-Interscience, New York, 1970.

[73] S. G. Chappell, "Automatic Test Generation for Asynchronous Digital Circuits,"*The Bell System Technical J.*, Vol. 53, pp. 1477-1503, Feb. 1974.

[74] A. Chatterjee and J. A. Abraham, "Test Generation for Arithmetic Units by Graph Labelling,"*Fault-Tolerant Computing Symp. (FTCS-17) Digest of Papers*, pp. 284-289, July 1987.

[75] A. Chatterjee and J. A. Abraham, "NCUBE: An Automatic Test Generation Program for Iterative Logic Arrays," *Proc. Int. Conf. CAD (ICCAD-88)*, Santa Clara, CA, November 1988, pp. 428-431.

[76] C. L. Chen and M. W. Du, "Multiple Stuck-Fault Detection and Location in Multivalued Linear Circuits,"*IEEE Trans. Computers*, Vol. C-35, pp. 1068-1071, December 1986.

[77] H. H. Chen, R. G. Mathews, and J. A. Newkirk, "Test Generation for MOS Circuits," *International Test Conference*, Philadelphia, PA, Oct. 1984, pp. 70-79.

[78] H. H. Chen, R. G. Mathews, and J. A. Newkirk, "An Algorithm to Generate Tests for MOS Circuits at the Switch Level," *Proc. Int. Test Conf.*, Philadelphia, PA, November 1985, pp. 304-312.

[79] K. T. Cheng, "A Simulation-Based Directed-Search Method for Test Generation," *Ph.D. Dissertation*, University of California, Berkeley, CA, June 1988.

[80] K. T. Cheng and V. D. Agrawal, "A Simulation-Based Directed-Search Method for Test Generation," *Proc. Int. Conf. Comp. Des. (ICCD-87)*, Rye Brook, NY, October 1987, pp. 48-51.

[81] K. T. Cheng and V. D. Agrawal, "Concurrent Test Generation and Design for Testability," *Proc. Int. Symp. Circ. and Syst. (ISCAS)*, Portland, Oregon, May 1989.

[82] K. T. Cheng and V. D. Agrawal, "An Economical Scan Design for Sequential Circuit Test Generation," *19th Fault-Tolerant Computing Symp. (FTCS-19) Digest of Papers*, Chicago, Illinois, June 1989.

[83] K. T. Cheng, V. D. Agrawal, and E. S. Kuh, "A Sequential Circuit Test Generator Using Threshold-Value Simulation,"*Fault-Tolerant Computing Symposium (FTCS-18) Digest of Papers*, pp. 24-29, June 1988.

[84] W. T. Cheng, "Split Circuit Model for Test Generation," *Proc. 25th Des. Auto. Conf.*, Anaheim, CA, June 1988, pp. 96-101.

[85] W. T. Cheng, "The BACK Algorithm for Sequential Test Generation," *Proc. IEEE Int. Conf. Comp. Design (ICCD-88)*, Rye Brook, NY, October 1988, pp. 66-69.

[86] W. T. Cheng and J. H. Patel, "A Minimum Test Set for Multiple-Fault Detection in Ripple Carry Adders," *Proc. Int. Conf. Comp. Des. (ICCD)*, Port Chester, NY, October 1985, pp. 435-438.

[87] W. T. Cheng and J. H. Patel, "A Shortest Length Test Sequence for Sequential-Fault Detection in Ripple Carry Adders," *Proc. Int. Conf. on CAD (ICCAD)*, Santa Clara, CA, November 1985, pp. 71-73.

[88] W. T. Cheng and J. H. Patel, "Testing in Two-dimensional Iterative Logic Arrays,"*Fault-Tolerant Comp. Symp. (FTCS-16) Digest of Papers*, pp. 76-81, July 1986.

[89] W. T. Cheng and J. H. Patel, "Testing in Two-Dimensional Iterative Logic Arrays,"*Comput. Math. Applic.*, Vol. 13, No. 5/6, 1987.

[90] W. T. Cheng and J. H. Patel, "A Minimum Test Set for Multiple Fault Detection in Ripple Carry Adders,"*IEEE Trans. Computers*, Vol. C-36, pp. 891-895, July 1987.

[91] K. W. Chiang and Z. G. Vranesic, "Test Generation for MOS Complex Gate Networks,"*Fault-Tolerant Comp. Symp. (FTCS-12) Digest of Papers*, pp. 149-157, June 1982.

[92] K. W. Chiang and Z. G. Vranesic, "On Fault Detection in CMOS Logic Networks," *Proc. 20th Des. Auto. Conf.*, Miami Beach, FL, June 1983,

pp. 50-56.

[93] R. J. Clarke, P. Arya, R. J. Potter, G. J. Smith, and I. D. Smith, "Hierarchical Design Verification and Incremental Test of a 100,000 Transistor Integrated Circuit," *Proc. Custom Integrated Circuit Conf.*, Rochester, NY, May 1984, pp. 601-605.

[94] B. Courtois, "Analytical Testing of Data Processing Sections of Integrated CPUs," *Int. Test Conf. Digest of Papers*, Philadelphia, PA, October 1981, pp. 21-28.

[95] H. Cox and J. Rajski, "A Method of Fault Analysis for Test Generation and Fault Diagnosis,"*IEEE Trans. CAD*, Vol. 7, pp. 813-833, July 1988.

[96] H. Cox and J. Rajski, "Stuck-Open and Transition Fault Testing in CMOS Complex Gates," *Proc. Int. Test Conf.*, Washington, DC, September 1988, pp. 688-694.

[97] M. Cutler, S. Y. H. Su, and M. Wang, "Test Generation by Critical Backtracing with Time Reducing Heuristics," *Proc. Int. Test Conf.*, Washington, DC, September 1987, pp. 1035-1042.

[98] W. Daehn, "A Unified Treatment of PLA Faults by Boolean Differences," *Proc. 23rd Des. Auto. Conf.*, Las Vegas, Nevada, June 1986, pp. 334-338.

[99] R. I. Damper and N. Burgess, "MOS Test Pattern Generation Using Path Algebras,"*IEEE Trans. Computers*, Vol. C-36, pp. 1123-1128, September 1987.

[100] S. R. Das, W. B. Jones, Z. Chen, A. K. Nath, and T. T. Lee, "Fault Location in Combinational Logic Networks by Multistage Binary Tree Classifier," *Proc. Int. Conf. Circ. Computers (ICCC'82)*, New York, NY, September 1982, pp. 624-628.

[101] R. David and P. Thevenod-Fosse, "Minimal Detecting Transition Sequences: Application to Random Testing,"*IEEE Trans. Computers*, Vol. C-29, pp. 514-518, June 1980.

[102] R. David and P. Thevenod-Fosse, "Random Testing of Intermittent Faults in Digital Circuits,"*Fault-Tolerant Comp. Symp. (FTCS-10) Digest of Papers*, pp. 182-184, October 1980.

[103] T. A. Davis, R. P. Kunda, and W. K. Fuchs, "Testing of Bit-Serial Multipliers," *Proc. Int. Conf. Comp. Des. (ICCD)*, Port Chester, NY, October 1985, pp. 430-434.

[104] C. Delorme, P. Roux, L. D. D'Archimbaud, N. Giambiasi, R. L'Bath, B.

MacGee, and R. Charroppin, "A Functional Partitioning Expert System for Test Sequences Generation," *Proc. 22nd Des. Auto. Conf.*, Las Vegas, Nevada, June 1985, pp. 820-824.

[105] J. DiGiacomo, *VLSI Handbook*, McGraw-Hill, New York, 1989.

[106] F. Distante and V. Piuri, "Optimum Behavioral Test Procedure for VLSI Devices: A Simulated Annealing Approach," *Proc. Int. Conf. Comp. Des. (ICCD)*, Port Chester, NY, October 1986, pp. 31-35.

[107] M. El-Lithy and R. Husson, "Bit-Sliced Microprocessors Testing – A Case Study,"*Fault-Tolerant Comp. Symp. (FTCS-10) Digest of Papers*, pp. 126-128, October 1980.

[108] Y. M. El-ziq, "A New Test Pattern Generation System," *Proc. 17th Des. Auto. Conf.*, Minneapolis, MN, June 1980, pp. 62-68.

[109] Y. M. El-ziq, "Automatic Test Generation for Stuck-Open Faults in CMOS VLSI," *Proc. 18th Des. Auto. Conf.*, Nashville, TN, June 1981, pp. 347-354.

[110] Y. M. El-ziq, H. H. Butt, and A. K. Bhatt, "An Automatic Test Pattern Generation Machine," *Proc. Int. Conf. CAD (ICCAD-84)*, Santa Clara, CA, November 1984, pp. 257-259.

[111] Y. M. El-ziq and R. J. Cloutier, "Functional-Level Test Generation for Stuck-Open Faults in CMOS VLSI," *Int. Test Conference Digest of Papers*, Oct. 1981, pp. 536-546.

[112] Y. M. El-ziq and S. Y. H. Su, "Fault Diagnosis of MOS Combinational Networks,"*IEEE Trans. Computers*, Vol. C-31, pp. 129-139, February 1982.

[113] P. P. Fasang, "A Fault Detection and Isolation Technique for Microcomputers," *Int. Test Conf., Digest of Papers*, Nov. 1982, pp. 214-219.

[114] X. Fedi and R. David, "Experimental Results from Random Testing of Microprocessors,"*Fault-Tolerant Comp. Symp. (FTCS-14) Digest of Papers*, pp. 225-230, June 1984.

[115] A. Feizi and D. Radhakrishnan, "Multiple Output Pass Networks: Design and Testing," *Proc. Int. Test Conf.*, Philadelphia, PA, November 1985, pp. 907-911.

[116] F. J. Ferguson and J. P. Shen, "Multiple-Fault Test Sets for MOS Complex Gates," *Proc. Int. Conf. on CAD (ICCAD)*, Santa Clara, CA, November 1985, pp. 36-38.

[117] D. Florcik and D. Low, "Simulation Pattern Capturing System for Design

Verification Using a Dynamic High Speed Functional Tester," *Proc. Int. Test Conf.*, Philadelphia, PA, October 1983, pp. 122-127.

[118] S. Freeman, "The F-Path Method of Test Generation for Datapath Logic," *Proc. Custom Integrated Circ. Conf.*, Portland, OR, May 1987, pp. 72-77.

[119] J. F. Frenzel and P. N. Marinos, "Functional Testing of Microprocessors in a User Environment," *Fault-Tolerant Comput. Symp. (FTCS-14), Digest of Papers*, pp. 219-224, June 1984.

[120] A. D. Friedman and P. R. Menon, *Fault Detection in Digital Circuits*, Prentice-Hall, Englewood Cliffs, NJ, 1971.

[121] A. Fuentes, R. David, and B. Courtois, "Random Testing Versus Deterministic Testing of RAMs," *Fault-Tolerant Comp. Symp. (FTCS-16) Digest of Papers*, pp. 266-271, July 1986.

[122] T. Fujieda and N. Arai, "Considerations of the Testing of RAMs with Dual Ports," *Proc. Int. Test Conf.*, Philadelphia, PA, November 1985, pp. 456-461.

[123] H. Fujiwara, "On Closedness and Test Complexity of Logic Circuits," *IEEE Trans. Computers*, Vol. C-30, pp. 556-562, August 1981.

[124] H. Fujiwara, *Logic Testing and Design for Testability*, MIT Press, Cambridge, MA, 1985.

[125] H. Fujiwara, K. Kinoshita, and H. Ozaki, "Universal Test Sets for Programmable Logic Arrays," *Fault-Tolerant Comp. Symp. (FTCS-10) Digest of Papers*, pp. 137-142, October 1980.

[126] H. Fujiwara and T. Shimono, "On the acceleration of test generation algorithms," *IEEE Trans. Comp.*, Vol. C-32, pp. 1137-1144, Dec. 1983.

[127] H. Fujiwara and S. Toida, "The Complexity of Fault Detection Problem for Combinational Logic Circuits," *IEEE Trans. Computers*, Vol. C-31, pp. 555-560, June 1982.

[128] S. Funatsu and H. Terai, "An Automatic Test-Generation System for Large Digital Circuits," *IEEE Design & Test of Computers*, Vol. 2, pp. 54-60, October 1985.

[129] R. K. Gaede, M. R. Mercer, K. M. Butler, and D. E. Ross, "CATAPULT: Concurrent Automatic Testing Allowing Parallelization and Using Limited Topology," *Proc. 25th Des. Auto. Conf.*, Anaheim, CA, June 1988, pp. 597-600.

[130] N. Giambiasi, L'Bath, B. MacGee, C. Delorme, P. Roux, and L. D.

D'Archimbaud, "An Adaptive and Evolutive Tool for Test Generation Process Based on Frames and Demons,"*Fault-Tolerant Comp. Symp. (FTCS-15) Digest of Papers*, pp. 50-57, June 1985.

[131] G. Giles and C. Hunter, "A Methodology for Testing Content Addressable Memories," *Proc. Int. Test Conf.*, Philadelphia, PA, November 1985, pp. 471-474.

[132] C. T. Glover and M. R. Mercer, "A Method of Delay Fault Test Generation," *Proc. 25th Des. Auto. Conf.*, Anaheim, CA, June 1988, pp. 90-95.

[133] P. Goel, "Test Generation Costs Analysis and Projections," *Proc. 17th Design Automation Conference*, Minneapolis, MN, June 1980, pp. 77-84.

[134] P. Goel, "An Implicit Enumeration Algorithm to Generate Tests for Combinational Logic Circuits,"*IEEE Trans. Computers*, Vol. C-30, pp. 215-222, March 1981. Also *ault-Tolerant Comp. Symp. (FTCS-10) Digest of Papers*, Kyoto, Japan, October 1980, pp. 145-151.

[135] P. Goel and B. C. Rosales, "PODEM-X: An Automatic Test Generation System for VLSI Logic Structures," *Proc. 18th Des. Auto. Conf.*, Nashville, TN, June 1981, pp. 260-268.

[136] A. V. Goldberg and K. J. Lieberherr, "Efficient Test Generation Algorithms," *Proc. Int. Test Conf.*, Philadelphia, PA, November 1985, pp. 508-516.

[137] J. Grason and A. W. Nagle, "Digital Test Generation and Design for Testability," *Proc. 17th Design Auto. Conf.*, Minneapolis, MN, June 1980, pp. 175-189.

[138] R. K. Gulati and D. K. Goel, "An Efficient Compaction Algorithm for Test Vectors of Microprocessors and Microcontrollers," *Proc. Int. Conf. CAD (ICCAD-88)*, Santa Clara, CA, November 1988, pp. 378-381.

[139] N. L. Gunther and W. C. Carter, "Remarks on the Probability of Detecting Faults,"*Fault-Tolerant Comp. Symp. (FTCS-10) Digest of Papers*, pp. 213-215, October 1980.

[140] G. Gupta and N. K. Jha, "A Universal Test Set for CMOS Circuits,"*IEEE Trans. CAD*, Vol. 7, pp. 590-597, May 1988.

[141] I. S. Gupta, "Index Vector Testing of Combinational Circuits," *Proc. Int. Test Conf.*, Washington, DC, September 1987, pp. 1108-1112.

[142] T. Hayashi, K. Hatayama, K. Sato, and T. Natabe, "A Delay Test Generator for Logic LSI,"*Fault-Tolerant Comput. Symp. (FTCS-14), Digest of Papers*, pp. 146-149, June 1984.

[143] J. P. Hayes, "Testing Memories for Single-Cell Pattern-Sensitive Faults,"*IEEE Trans. Computers*, Vol. C-29, pp. 249-254, March 1980.

[144] B. J. Heard, R. N. Sheshadri, R. B. David, and A. G. Sammuli, "Automatic Test Pattern Generation for Asynchronous Networks," *Proc. Int. Test Conf.*, Philadelphia, PA, October 1984, pp. 63-69.

[145] F. C. Hennie, "Fault Detecting Experiments for Sequential Circuits," *Proc. 5th Ann. Symp. Sw. Cir. Theory and Logical Des.*,", November 1964, pp. 95-110.

[146] F. J. Hill and B. Huey, "A Design Language Based Approach to Test Sequence Generation,"*Computer*, pp. 28-33, June 1977.

[147] C. Hinchcliff, "Simplified Microprocessor Test Generation," *Int. Test Conf. Digest of Papers*, Philadelphia, PA, November 1982, pp. 176-180.

[148] F. Hirose, K. Takayama, and N. Kawato, "A Method to Generate Tests for Combinational Logic Circuits Using an Ultra-High-Speed Logic Simulator," *Proc. Int. Test Conf.*, Washington, DC, September 1988, pp. 102-107.

[149] H. Hofestadt and M. Gerner, "Qualitative Testability Analysis and Hierarchical Test Pattern Generation – A New Approach to Design for Testability," *Proc. Int. Test Conf.*, Washington, DC, September 1987, pp. 538-546.

[150] S. J. Hong, "Existence Algorithms for Synchronizing/Distinguishing Sequences,"*IEEE Trans. Computers*, Vol. C-30, pp. 234-237, March 1981.

[151] S. J. Hong and D. L. Ostapko, "A Simple Procedure to Generate Optimum Test Patterns for Parity Logic Networks,"*IEEE Trans. Computers*, Vol. C-30, pp. 356-358, May 1981.

[152] E. P. Hsieh, G. R. Putzolu, and C. J. Tan, "A Test Pattern Generation System for Sequential Logic Circuits,"*Fault-Tolerant Comp. Symp. (FTCS-11) Digest of Papers*, pp. 230-232, June 1981.

[153] R. V. Hudli, W. Ke, and S. C. Seth, "Structural Profile of Benchmark Circuits Relating to the Test Generation Problems," in *VLSI Design: Proc. Second Int. Workshop on VLSI Design*, R. Apte, V. D. Agrawal, and A. Prabhakar, Eds., Tata McGraw-Hill, New Delhi, India, 1988, pp. 438-447.

[154] L. M. Huisman, "Random Pattern Testing of Logic Surrounding Embedded RAMS using Perturbation Analysis," *Proc. Int. Conf. Comp. Des. (ICCD)*, Port Chester, NY, October 1986, pp. 20-23.

[155] A. C. Hung and F. C. Wang, "A Method for Test Generation Directly from Testability Analysis," *Proc. Int. Test Conf.*, Philadelphia, PA, November 1985, pp. 62-78.

[156] A. Hunger and A. Gaertner, "Functional Characterization of Microprocessors," *Proc. Int. Test Conf.*, Philadelphia, PA, October 1984, pp. 794-803.

[157] K. S. Hwang and M. R. Mercer, "Informed Test Generation Guidance Using Partially Specified Fanout Constraints," *Proc. Int. Test Conf.*, Washington, DC, September 1986, pp. 113-120.

[158] K. S. Hwang and M. R. Mercer, "Derivation and Refinement of Fan-out Constraints to Generate Tests in Combinational Logic Circuits,"*IEEE Trans. on CAD*, Vol. CAD-5, pp. 564-572, October 1986. Also *Proc. Int. Conf. on CAD (ICCAD)*, Santa Clara, CA, November 1985, pp. 10-12.

[159] O. H. Ibarra and S. K. Sahni, "Polynomially Complete Fault Detection Problems,"*IEEE Trans. Comp.*, Vol. C-24, pp. 242-249, March 1975.

[160] K. Itazaki and K. Kinishita, "Test Pattern Generation for Circuits with Three-State Modules by Improved Z-Algorithm," *Proc. Int. Test Conf.*, Washington, DC, September 1986, pp. 105-112.

[161] N. Itazaki and K. Kinoshita, "Algorithmic Generation of Test Patterns for Circuits with Tri-state Modules,"*Fault-Tolerant Comp. Symp. (FTCS-16) Digest of Papers*, pp. 64-69, July 1986.

[162] V. S. Iyengar, B. K. Rosen, and I. Spillinger, "Delay Test Generation 1: Concepts and Coverage Metrics," *Proc. Int. Test Conf.*, Washington, DC, September 1988, pp. 857-866.

[163] V. S. Iyengar, B. K. Rosen, and I. Spillinger, "Delay Test Generation 2: Algebra and Algorithms," *Proc. Int. Test Conf.*, Washington, DC, September 1988, pp. 867-876.

[164] D. M. Jacobson, "A Fast, Probabilistic Algorithm for Functional Testing of Random Access Memory Systems," *Proc. Int. Test Conf.*, Philadelphia, PA, November 1985, pp. 169-177.

[165] S. K. Jain and V. D. Agrawal, "Test Generation for MOS Circuits Using D-Algorithm," *Proc. ACM/IEEE 20th Design Automation Conf.*, Miami Beach, FL, June 1983, pp. 64-70.

[166] S. K. Jain and V. D. Agrawal, "Modeling and Test Generation Algorithms for MOS Circuits,"*IEEE Trans. Computers*, Vol. C-34, pp. 426-433, May 1985. Also see, Correction, in *IEEE Trans. Computers*, Vol. C-34, p. 680, July 1985.

[167] S. K. Jain and A. K. Susskind, "Test Strategy for Microprocessors," *Proc. 20th Des. Auto. Conf.*, Miami Beach, FL, June 1983, pp. 703-708.

[168] N. K. Jha, "Testing of Cascode Voltage Switch Parity Trees," in *VLSI Design: Proc. Second Int. Workshop on VLSI Design*, R. Apte, V. D. Agrawal, and A. Prabhakar, Eds., Tata McGraw-Hill, New Delhi, India, 1988, pp. 267-276.

[169] M. Johansson, "The GENESYS – Algorithm for ATPG without Fault Simulation," *Proc. Int. Test Conf.*, Philadelphia, PA, October 1983, pp. 333-337.

[170] P. D. Jong and A. J. VandeGoor, "Test Pattern Generation for API Faults in RAM,"*IEEE Trans. Computers*, Vol. 37, pp. 1426-1428, November 1988.

[171] J. Y. Jou, "A Testable PLA Design with Low Overhead and Ease of Test Generation," *Proc. IEEE Int. Conf. Comp. Design (ICCD-88)*, Rye Brook, NY, October 1988, pp. 450-453.

[172] M. Karpovsky, "Universal Tests Detecting Input/Output Faults in Almost All Devices," *Int. Test Conf. Digest of Papers*, Philadelphia, PA, November 1982, pp. 52-57.

[173] M. Karpovsky, "Universal Tests for Detection of Input/Output Stuck-at and Bridging Faults,"*IEEE Trans. Computers*, Vol. C-32, pp. 1194-1198, Dec. 1983.

[174] M. Karpovsky and L. Levitin, "Detection and Identification of Input/Output Stuck-at and Bridging Faults in Combinational and Sequential VLSI Networks by Universal Tests,"*Integration, the VLSI Journal*, Vol. 1, pp. 211-232, October 1983.

[175] M. Karpovsky and S. Y. H. Su, "Detection and Location of Input and Feedback Bridging Faults among Input and Output Lines,"*IEEE Trans. Computers*, Vol. C-29, pp. 523-527, June 1980. Also see correction in *IEEE Trans. Comp.*, Vol. C-30, p. 86, Jan. 1981.

[176] M. Kawai, H. Shibano, S. Funatsu, S. Kato, T. Kurobe, K. Ookawa, and T. Sasaki, "A High Level Test Pattern Generation Algorithm," *Proc. Int. Test Conf.*, Philadelphia, PA, October 1983, pp. 346-352.

[177] R. Khorram, "Functional Test Pattern Generation for Integrated Circuits," *Proc. Int. Test Conf.*, Philadelphia, PA, October 1984, pp. 246-249.

[178] G. Kildiran and P. N. Marinos, "Functional Testing of Microprocessor-Like Architectures," *Proc. Int. Test Conf.*, Washington, D.C., September

1986, pp. 913-920.

[179] K. Kinoshita and H. Fujiwara, *Fault Diagnosis of Digital Circuits*, Vol. 1, Kogaku-Tosho, Tokyo, Japan, 1983. In Japanese.

[180] T. Kirkland and M. R. Mercer, "A Topological Search Algorithm for ATPG," *Proc. 24th Des. Auto. Conf.*, Miami Beach, FL, June 1987, pp. 502-508.

[181] T. Kirkland and M. R. Mercer, "Algorithms for Automatic Test Pattern Generation,"*IEEE Design & Test of Computers*, Vol. 5, pp. 43-55, June 1988.

[182] H. P. Klug, "Microprocessor Testing by Instruction Sequences Derived from Random Patterns," *Proc. Int. Test Conf.*, Washington, DC, September 1988, pp. 73-80.

[183] T. J. Knips and D. J. Malone, "Designing Characterization Tests for Bipolar Array Performance Verification," *Proc. Int. Test Conf.*, Washington, D.C., September 1986, pp. 840-845.

[184] K. Kobayashi, "Functional Test Data Generation for Hardware Design Verification," *Proc. Int. Test Conf.*, Washington, DC, September 1987, pp. 547-552.

[185] B. Koenemann, J. Ducklow, N. Lanners, and T. Vriezen, "Computer Aided Test for VLSI," *Proc. Custom Integrated Circuits Conf.*, Rochester, NY, May 1984, pp. 172-175.

[186] G. A. Kramer, "Employing Massive Parallelism in Digital ATPG Algorithms," *Proc. Int. Test Conf.*, Philadelphia, PA, October 1983, pp. 108-114.

[187] B. Krishnamurthy, "Hierarchical Test Generation: Can AI Help?," *Proc. Int. Test Conf.*, Washington, DC, September 1987, pp. 694-700.

[188] B. Krishnamurthy and S. B. Akers, "On the Complexity of Estimating the Size of a Test Set,"*IEEE Trans. Comput.*, Vol. C-33, pp. 750-753, Aug. 1984.

[189] G. Krueger, "Automatic Generation of Self-Test Programs – A New Feature of the MIMOLA Design System," *Proc. 23rd Des. Auto. Conf.*, Las Vegas, Nevada, June 1986, pp. 378-384.

[190] G. Krueger, "Tools for Hierarchical Test Generation," *Proc. Int. Conf. CAD (ICCAD-88)*, Santa Clara, CA, November 1988, pp. 420-423.

[191] J. R. Kuban and J. E. Salick, "Testing Approaches in the MC68020,"*VLSI Design*, Vol. V, p. 22, November 1984.

[192] H. Kubo, "A Procedure for Generating Test Sequences to Detect Sequential Circuit Failures,"*NEC Res. & Dev.*, No. 12, pp. 69-78, October 1968.

[193] S. Kundu and S. M. Reddy, "Robust Tests for Parity Trees," *Proc. Int. Test Conf.*, Washington, DC, September 1988, pp. 680-687.

[194] M. Ladjadj and J. F. McDonald, "Benchmark Runs of the Subscripted D-Algorithm with Observation Path Mergers on the Brglez-Fujiwara Circuits," *Proc. 24th Des. Auto. Conf.*, Miami Beach, FL, June 1987, pp. 509-515.

[195] M. Ladjadj, J. F. McDonald, D. H. Ho, and W. Murray Jr., "Use of the Subscripted DALG in Submodule Testing with Applications in Cellular Arrays," *Proc. 23rd Des. Auto. Conf.*, Las Vegas, Nevada, June 1986, pp. 346-353.

[196] K. W. Lai, "Test Program Compiler – A High Level Test Program Specification Language," *Proc. Int. Conf. CAD (ICCAD-83)*, Santa Clara, CA, September 1983, pp. 30-31.

[197] K. W. Lai and D. P. Siewiorek, "Functional Testing of Digital Systems," *Proc. 20th Des. Auto. Conf.*, Miami Beach, FL, June 1983, pp. 207-213.

[198] P. K. Lala, *Fault Tolerant & Fault Testable Hardware Design*, Prentice-Hall International, London, UK, 1985.

[199] P. Lamoureux and V. K. Agarwal, "Non-Stuck-At Fault Detection in nMOS Circuits by Region Analysis," *Proc. Int. Test Conf.*, Philadelphia, PA, October 1983, pp. 129-137.

[200] J. D. Lesser and J. J. Shedletsky, "An Experimental Delay Test Generator for LSI Logic,"*IEEE Trans. Computers*, Vol. C-29, pp. 235-248, March 1980.

[201] Y. Levendel and P. R. Menon, "The *-Algorithm: Critical Traces for Functions and CHDL Constructs,"*Fault-Tolerant Comp. Symp. (FTCS-13) Digest of Papers*, pp. 90-97, June 1983.

[202] Y. H. Levendel and P. R. Menon, "Test Generation Algorithms for Nonprocedural Computer Hardware Description Languages,"*Fault-Tolerant Comp. Symp. (FTCS-11) Digest of Papers*, pp. 200-205, June 1981.

[203] M. W. Levi, "Opens Tests for CMOS,"*IEEE J. Solid State Circ.*, Vol. SC-22, p. 129, February 1987.

[204] W. N. Li, S. M. Reddy, and S. K. Sahni, "On Path Selection in Combinational Logic Circuits,"*IEEE Trans. on CAD*, Vol. 8, pp. 56-63, January

1989.

[205] C. Liaw, S. Y. H. Su, and Y. K. Malaiya, "Test Generation for Delay Faults Using Stuck-at-Fault Test Set," *Test Conference Digest of Papers*, Philadelphia, PA, November 1980, pp. 167-175.

[206] C. C. Liaw, S. Y. H. Su, and Y. K. Malaiya, "State Diagram Approach for Functional Testing of Control Section," *Int. Test Conf. Digest of Papers*, Philadelphia, PA, October 1981, pp. 433-446.

[207] C. S. Lin and H. F. Ho, "Automatic Functional Test Program Generation for Microprocessors," *Proc. 25th Des. Auto. Conf.*, Anaheim, CA, June 1988, pp. 605-608.

[208] G. K. Lin and P. R. Menon, "Totally Preset Checking Experiments for Sequential Machines,"*IEEE Trans. Computers*, Vol. C-32, pp. 101-108, February 1983.

[209] M. G. Lin and K. Rose, "Applying Test Theory to VLSI Testing," *Int. Test Conf. Digest of Papers*, Philadelphia, PA, November 1982, pp. 580-586.

[210] T. Lin and S. Y. H. Su, "Functional Test Generation of Digital LSI/VLSI Systems Using Machine Symbolic Execution Technique," *Proc. Int. Test Conf.*, Philadelphia, PA, October 1984, pp. 660-668.

[211] T. Lin and S. Y. H. Su, "The S-Algorithm: A Promising Solution for Systematic Functional Test Generation,"*IEEE Trans. on CAD*, Vol. CAD-4, pp. 250-263 Also *Proc. Int. Conf. CAD (ICCAD-84)*, Santa Clara, CA, Nov. 1984, pp. 134-136, July 1985.

[212] T. Lin and S. Y. H. Su, "VLSI Functional Test Pattern Generation – A Design and Implementation," *Proc. Int. Test Conf.*, Philadelphia, PA, November 1985, pp. 922-929.

[213] A. Lioy, "Adaptive Backtrace and Dynamic Partitioning Enhance ATPG," *Proc. IEEE Int. Conf. Comp. Design (ICCD-88)*, Rye Brook, NY, October 1988, pp. 62-65.

[214] R. Lisanke, F. Brglez, A. deGeus, and D. Gregory, "Testability-Driven Random Pattern Generation,"*IEEE Trans. CAD*, Vol. CAD-6, pp. 1082-1087, November 1987. Also *Proc. Int. Conf. CAD (ICCAD-86)*, Santa Clara, CA, Nov., 1986, pp. 144-147.

[215] *Testing and Diagnosis of VLSI and ULSI*, Kluwer Academic Publishers, Boston, MA, 1988.

[216] P. K. Lui and J. C. Muzio, "Spectral Signature Testing of Multiple

Stuck-at Faults in Irredundant Combinational Networks,"*IEEE Trans. Computers*, Vol. C-35, pp. 1088-1092, December 1986.

[217] H. K. T. Ma, S. Devadas, A. R. Newton, and A. Sangiovanni-Vincentelli, "Test Generation for Sequential Circuits,"*IEEE Trans. CAD*, Vol. 7, pp. 1081-1093, October 1988.

[218] R. H. Macmillan and M. R. Bentley, "An Efficient Test Vector Generation and Reduction Method for an LSI Digital Filter Circuit Using an Adaptive Search Technique," *Int. Test Conf. Digest of Papers*, Philadelphia, PA, November 1982, pp. 601-605.

[219] S. R. Makar and E. J. McCluskey, "On the Testing of Multiplexers," *Proc. Int. Test Conf.*, Washington, DC, September 1988, pp. 669-679.

[220] S. Mallela and S. Wu, "A Sequential Circuit Test Generation System," *Proc. Int. Test Conf.*, Philadelphia, PA, November 1985, pp. 57-61.

[221] M. Mannan, "Instability – A CAD Dilemma," *Proc. Int. Test Conf.*, Washington, D.C., September 1986, pp. 637-643.

[222] W. Mao and M. D. Ciletti, "DYTEST: A Self-Learning Algorithm Using Dynamic Testability Measures to Accelerate Test Generation," *Proc. 25th Des. Auto. Conf.*, Anaheim, CA, June 1988, pp. 591-596.

[223] W. Mao and X. Ling, "Robust Test Generation Algorithm for Stuck-Open Fault in CMOS Circuits," *Proc. 23rd Des. Aut. Conf.*, Las Vegas, Nevada, June 1986, pp. 236-242.

[224] M. Marinescu, "Simple and Efficient Algorithms for Functional RAM Testing," *Int. Test Conf. Digest of Papers*, Philadelphia, PA, November 1982, pp. 236-239.

[225] R. Marlett, "An Effective Test Generation System for Sequential Circuits," *23rd Des. Aut. Conf.*, Las Vegas, Nevada, June 1986, pp. 250-256.

[226] R. Marlett, "Automated Test Generation for Integrated Circuits,"*VLSI Systems Design*, Vol. VII, pp. 68-73, July 1986.

[227] R. Marlett and S. R. Pollock, "Guaranteeing ASIC Testability,"*VLSI Systems Design*, Vol. IX, pp. 70-76, August 1988.

[228] W. P. Marnane and W. R. Moore, "Testing of VLSI Regular Arrays," *Proc. IEEE Int. Conf. Comp. Design (ICCD-88)*, Rye Brook, NY, October 1988, pp. 145-148.

[229] W. P. Marnane, W. R. Moore, H. M. Yassine, E. Gautrin, N. Burgess, and A. P. H. McCabe, "Testing Bit-Level Systolic Arrays," *Proc. Int. Test Conf.*, Washington, DC, September 1987, pp. 906-914.

[230] C. Maunder, "HITEST Test Generation System – Interfaces," *Proc. Int. Test Conf.*, Philadelphia, PA, October 1983, pp. 324-332.

[231] P. Mazumder, J. H. Patel, and W. K. Fuchs, "Design and Algorithms for Parallel Testing of Random Access and Content Addressable Memories," *Proc. 24th Des. Auto. Conf.*, Miami Beach, FL, June 1987, pp. 688-694.

[232] W. H. McAnney, J. Savir, and S. R. Vecchio, "Random Pattern Testing for Address-line Faults in an Embedded Multiport Memory," *Proc. Int. Test Conf.*, Philadelphia, PA, November 1985, pp. 106-114.

[233] W. H. McAnney, J. Savir, and S. R. Vecchio, "Random Pattern Testing for Data-line Faults in an Embedded Multiport Memory," *Proc. Int. Test Conf.*, Philadelphia, PA, November 1985, pp. 100-105.

[234] E. J. McCluskey, "Verification Testing," *Proc. 19th Des. Auto. Conf.*, Las Vegas, Nevada, June 1982, pp. 495-500.

[235] J. F. McDonald and C. Benmehrez, "Test Set Reduction Using Subscripted D-Algorithm," *Proc. Int. Test Conf.*, Philadelphia, PA, October 1983, pp. 115-121.

[236] J. F. McDonald and C. Benmehrez, "Test Set Reduction Using the Subscripted D-Algorithm," *Proc. International Test Conference*, Philadelphia, PA, October 1983, pp. 115-121.

[237] M. R. Mercer, V. D. Agrawal, and C. M. Roman, "Test Generation for Highly Sequential Scan-Testable Circuits Through Logic Transformation," *Int. Test Conference Digest of Papers*, Philadelphia, PA, Oct. 1981, pp. 561-565.

[238] A. Miczo, "The Sequential ATPG: A Theoretical Limit," *Proc. Int. Test Conf.*, Philadelphia, PA, October 1983, pp. 143-147.

[239] A. Miczo, *Digital Logic Testing and Simulation*, Harper & Row, New York, 1986.

[240] T. Middleton, "Functional Test Vector Generation for Digital LSI/VLSI Devices," *Proc. Int. Test Conf.*, Philadelphia, PA, October 1983, pp. 682-691.

[241] P. J. Miller and G. E. Taylor, "Automatic Test Generation for VLSI," *Proc. Int. Conf. Circ. Computers (ICCC'82)*, New York, NY, September 1982, pp. 452-455.

[242] H. Miyamoto, K. Mashiko, Y. Morooka, K. Arimoto, M. Yamada, and T. Nakano, "Test Pattern Considerations for Fault Tolerant High Density DRAM," *Proc. Int. Test Conf.*, Philadelphia, PA, November 1985, pp.

451-455.

[243] A. Motohara, K. Nishimura, H. Fujiwara, and I. Shirakawa, "A Parallel Scheme for Test-Pattern Generation," *Proc. Int. Conf. CAD (ICCAD-86)*, Santa Clara, CA, November 1986, pp. 156-159.

[244] S. Mourad, "An Optimized ATPG," *Proc. 17th Des. Auto. Conf.*, Minneapolis, MN, June 1980, pp. 381-385.

[245] E. I. Muehldorf, "Test Pattern Generation as a Part of the Total Design Process," *Semiconductor Test Conf., Digest of Papers*, Cherry Hill, NJ, October, 1978, pp. 4-7.

[246] E. I. Muehldorf, G. P. Papp, and T. W. Williams, "Efficient Test Pattern Generation for Embedded PLAs," *Test Conference Digest of Papers*, Philadelphia, PA, November 1980, pp. 349-358.

[247] E. I. Muehldorf and T. W. Williams, "Analysis of the Switching Behavior of Combinatorial Logic Networks," *Int. Test Conf. Digest of Papers*, Philadelphia, PA, November 1982, pp. 379-390.

[248] M. Murakami, N. Shiraki, and K. Hirakawa, "Logic Verification and Test Generation for LSI Circuits," *Test Conference Digest of Papers*, Philadelphia, PA, November 1980, pp. 467-472.

[249] B. T. Murray and J. P. Hayes, "Hierarchical Test Generation Using Precomputed Tests for Modules," *Proc. Int. Test Conf.*, Washington, DC, September 1988, pp. 221-229.

[250] P. Muth, "A Nine-Valued Circuit Model for Test Generation,"*IEEE Trans. Computers*, Vol. C-25, pp. 630-636, June 1976.

[251] J. C. Muzio and D. M. Miller, "Spectral Techniques for Fault Detection,"*Fault-Tolerant Comp. Symp. (FTCS-12) Digest of Papers*, pp. 297-302, June 1982.

[252] S. Naito and M. Tsunoyama, "Fault Detection for Sequential Machines by Transition Tours,"*Fault-Tolerant Comp. Symp. (FTCS-11) Digest of Papers*, pp. 238-243, June 1981.

[253] A. S. Nale and S. Y. H. Su, "Testing a PLA without Augmentation," *Proc. Int. Test Conf.*, Washington, DC, September 1987, pp. 111-118.

[254] S. Nitta, K. Kawamura, and K. Hirabayashi, "Test Generation by Activation and Defect-Drive (TEGAD),"*Integration, The VLSI Journal*, Vol. 3, pp. 3-12, March 1985.

[255] S. E. Noujaim, R. T. Jerdonek, and S. J. Hong, "A Structured Approach to Test Vector Generation," *Proc. Int. Conf. Comp. Des. (ICCD'84)*, Port

Chester, NY, October 1984, pp. 757-762.

[256] T. Ogihara, S. Murai, Y. Takamatsu, and K. Kinoshita, "Test Generation for Scan Design Circuits with Tri-State Modules and Bidirectional Terminals," *Proc. 20th Des. Auto. Conf.*, Miami Beach, FL, June 1983, pp. 71-78.

[257] T. Ogihara and S. Saruyama, "Test Generation for Sequential Circuits Using Individual Initial Value Propagation," *Proc. Int. Conf. CAD (ICCAD-88)*, Santa Clara, CA, November 1988, pp. 424-427.

[258] T. Ogihara, S. Saruyama, and S. Murai, "PATEGE: An Automatic DC Parameter Test Generation System for Series Gated ECL Circuits," *Proc. 22nd Des. Auto. Conf.*, Las Vegas, Nevada, June 1985, pp. 212-218.

[259] C. A. Papachristou and N. B. Sahagal, "An Improved Method for Detecting Functional Faults in Semiconductor Random Access Memories," *IEEE Trans. Computers*, Vol. C-34, pp. 110-116, February 1985.

[260] E. S. Park and M. R. Mercer, "Robust and Nonrobust Tests for Path Delay Faults in a Combinational Circuit," *Proc. Int. Test Conf.*, Washington, DC, September 1987, pp. 1027-1034.

[261] K. P. Parker, "Adaptive Random Test Generation," *J. Des. Auto. and Fault-Tolerant Computing*, Vol. 1, pp. 62-83, October 1976.

[262] S. Patel and J. Patel, "Effectiveness of Heuristics Measures for Automatic Test Pattern Generation," *Proc. Des. Auto. Conf.*, Las Vegas, Nevada, June 1986, pp. 547-552.

[263] C. Paulson, "Classes of Diagnostic Tests," *Proc. 20th Des. Auto. Conf.*, Miami Beach, FL, June 1983, pp. 316-322.

[264] V. Pitchumani and S. Soman, "An Application of Unate Function Theory to Functional Testing," *Fault-Tolerant Comp. Symp. (FTCS-16) Digest of Papers*, pp. 70-75, July 1986.

[265] D. K. Pradhan, Ed., *Fault-Tolerant Computing Theory and Techniques*, I and II, Prentice-Hall, Englewood Cliffs, NJ, 1986.

[266] G. R. Putzolu and J. P. Roth, "A Heuristic Algorithm for the Testing of Asynchronous Circuits," *IEEE Trans. Computers*, Vol. C-20, pp. 639-647, June 1971.

[267] J. Rajski and H. Cox, "Stuck-Open Fault Testing in Large CMOS Networks by Dynamic Path Tracing," *Proc. Int. Conf. Comp. Des. (ICCD)*, Port Chester, NY, October 1986, pp. 252-255.

[268] J. Rajski and H. Cox, "A Method of Test Generation and Fault Diagnosis

in Very Large Combinational Circuits," *Proc. Int. Test Conf.*, Washington, DC, September 1987, pp. 932-943.

[269] J. Rajski and J. Tyszer, "Combinatorial Approach to Multiple Contact Faults Coverage in Programmable Logic Arrays,"*IEEE Trans. Computers*, Vol. C-34, pp. 549-553, June 1985.

[270] M. K. Reddy, S. M. Reddy, and P. Agrawal, "Transistor Level Test Generation for MOS Circuits," *Proc. 22nd Des. Auto. Conf.*, Las Vegas, Nevada, June 1985, pp. 825-828.

[271] S. M. Reddy, "Complete Test Sets for Logic Functions,"*IEEE Trans. Comp.*, Vol. C-22, pp. 1016-1020, Nov. 1973.

[272] S. M. Reddy, M. K. Reddy, and V. D. Agrawal, "Robust Tests for Stuck-Open Faults in CMOS Combinational Circuits,"*Fault-Tolerant Comput. Symp. (FTCS-14) Digest of Papers*, pp. 44-49, June 1984.

[273] E. Regener, "A Transition Sequence Generator for RAM Fault Detection,"*IEEE Trans. Computers*, Vol. 37, pp. 362-368, March 1988.

[274] C. Robach, P. Malecha, and G. Michel, "Computer Aided Testability Evaluation and Test Generation," *Proc. Int. Test Conf.*, Philadelphia, PA, October 1983, pp. 338-345.

[275] C. Robach and G. Saucier, "Application Oriented Microprocessor Test Method,"*Fault-Tolerant Comp. Symp. (FTCS-10) Digest of Papers*, pp. 121-125, October 1980.

[276] C. Robach and G. Saucier, "Microprocessor Functional Testing," *Test Conference Digest of Papers*, Philadelphia, PA, November 1980, pp. 433-443.

[277] G. D. Robinson, "HITEST – Intelligent Test Generation," *Proc. Int. Test Conf.*, Philadelphia, PA, October 1983, pp. 311-323.

[278] M. Robinson and J. Rajski, "An Algorithmic Branch and Bound Method for PLA Test Pattern Generation," *Proc. Int. Test Conf.*, Washington, DC, September 1988, pp. 784-795.

[279] S. H. Robinson and J. P. Shen, "Towards a Switch-Level Test Pattern Generation Program," *Proc. Int. Conf. on CAD (ICCAD)*, Santa Clara, CA, November 1985, pp. 39-41.

[280] B. C. Rosales and P. Goel, "Results from Application of a Commercial ATG System to Large-Scale Combinational Circuits," *Proc. Int. Symp. Circ. Syst. (ISCAS-85)*, Kyoto, Japan, June 1985, pp. 667-670.

[281] J. P. Roth, W. G. Bouricius, and P. R. Schneider, "Programmed

Algorithms to Compute Tests and to Detect and Distinguish Between Failures in Logic Circuits,''*IEEE Trans. Electronic Computers*, Vol. EC-16, pp. 567-580, October 1967.

[282] J. P. Roth, V. G. Oklobdzija, and J. F. Beetem, ''Test Generation for FET Switching Circuits,'' *Proc. Int. Test Conf.*, Philadelphia, PA, October 1984, pp. 59-62.

[283] R. K. Roy, T. M. Niermann, J. H. Patel, J. A. Abraham, and R. A. Saleh, ''Compaction of ATPG-Generated Test Sequences for Sequential Logic,'' *Proc. Int. Conf. CAD (ICCAD-88)*, Santa Clara, CA, November 1988, pp. 382-285.

[284] G. Russell and I. L. Sayers, *Advanced Simulation and Test Methodologies for VLSI Design*, Van Nostrand Reinhold, New York, 1989.

[285] J. Salick, B. Underwood, J. Kuban, and M. R. Mercer, ''An Automatic Test Pattern Generation Algorithm for PLAs,'' *Proc. Int. Conf. CAD (ICCAD-86)*, Santa Clara, CA, November 1986, pp. 152-155.

[286] K. K. Saluja and K. Kinoshita, ''Test Pattern Generation For API Faults In RAM,''*IEEE Transaction On Computers*, Vol. C-34, pp. 284-287, March 1985.

[287] K. K. Saluja, L. Shen, and S. Y. H. Su, ''A Simplified Algorithm for Testing Microprocessors,'' *Proc. Int. Test Conf.*, Philadelphia, PA, October 1983, pp. 668-675.

[288] K. K. Saluja, L. Shen, and S. Y. H. Su, ''A Simplified Algorithm for Testing Microprocessors,''*Comput. Math. Applic.*, Vol. 13, No. 5/6, pp. 431-441, 1987.

[289] E. F. Sarkany and W. S. Hart, ''Minimal Set of Patterns to Test RAM Components,'' *Proc. Int. Test Conf.*, Washington, DC, September 1987, pp. 759-764.

[290] T. Sasaki, S. Kato, N. Nomizu, and H. Tanaka, ''Logic Design Verification Using Automated Test Generation,'' *Proc. Int. Test Conf.*, Philadelphia, PA, October 1984, pp. 88-94.

[291] S. Sastry and M. A. Breuer, ''Detectability of CMOS Stuck-Open Faults Using Random and Pseudorandom Test Sequences,''*IEEE Trans. CAD*, Vol. 7, pp. 933-946, September 1988.

[292] G. Saucier and C. Bellon, ''CADOC : A System for Computed Aided Functional Test,'' *Proc. Int. Test Conf.*, Philadelphia, PA, October 1984, pp. 680-687.

[293] J. Savir, "Testing for Single Intermittent Failures in Combinational Circuits by Maximizing the Probability of Fault Detection,"*IEEE Trans. Computers*, Vol. C-29, pp. 410-416, May 1980.

[294] J. Savir, "Detection of Single Intermittent Faults in Sequential Circuits,"*IEEE Trans. Computers*, Vol. C-29, pp. 673-678, July 1980.

[295] J. Savir and P. H. Bardell, "On Random Pattern Test Length," *Proc. Int. Test Conf.*, Philadelphia, PA, October 1983, pp. 95-106.

[296] J. Savir, W. H. McAnney, and S. R. Vecchio, "Fault Propagation through Embedded Multiport Memories,"*IEEE Trans. Computers*, Vol. C-36, pp. 592-602, May 1987.

[297] J. T. Scanlon and W. K. Fuchs, "A Testing Strategy for Bit-Serial Arrays," *Proc. Int. Conf. CAD (ICCAD-86)*, Santa Clara, CA, November 1986, pp. 284-287.

[298] P. R. Schneider, "On the Necessity to Examine D-Chains in Diagnostic Test Generation,"*IBM J. Res. and Dev.*, Vol. 11, p. 114, January 1967.

[299] D. M. Schuler, E. G. Ulrich, T. E. Baker, and S. P. Bryant, "Random Test Generation Using Concurrent Logic Simulation," *Proc. 12th Des. Auto. Conf.*, 1975, pp. 261-267.

[300] M. H. Schulz and E. Auth, "Advanced Automatic Test Pattern Generation and Redundancy Identification Techniques,"*Fault-Tolerant Computing Symposium (FTCS-18) Digest of Papers*, pp. 30-35, June 1988.

[301] M. H. Schulz, E. Trischler, and T. M. Sarfert, "SOCRATES: A Highly Efficient Automatic Test Pattern Generation System," *Proc. Int. Test Conf.*, Washington, DC, September 1987, pp. 1016-1026. Also *IEEE Trans. CAD*, Vol. 7, pp. 126-137, January 1988.

[302] D. Sciuto and F. Lombardi, "On Functional Testing of Array Processors,"*IEEE Trans. Computers*, Vol. 37, pp. 1480-1484, November 1988.

[303] P. Seetharamaiah and V. R. Murthy, "Tabular Mechanism for Flexible Testing of Microprocessors," *Proc. Int. Test Conf.*, Washington, DC, September 1986, pp. 394-407.

[304] F. F. Sellers, M. Y. Hsiao, and L. W. Bearnson, "Analyzing Errors with Boolean Difference,"*IEEE Trans. Computers*, Vol. C-17, pp. 676-683, July 1968.

[305] S. C. Seth and V. D. Agrawal, "Statistical Design Verification,"*Fault-Tolerant Comp. Symp. (FTCS-12) Digest of Papers*, pp. 393-399, June

1982.

[306] S. Shalem, "Functional Testing of the NS32332 Microprocessor," *Proc. Int. Test Conf.*, Washington, D.C., September 1986, pp. 552-560.

[307] R. Sharma and D. Radhakrishnan, "Test Derivation for CMOS Iterative Logic Arrays," *Proc. Custom Integrated Circ. Conf.*, Portland, OR, May 1985, pp. 315-318.

[308] L. Shen and S. Y. H. Su, "A Functional Testing Method for Microprocessors,"*Fault-Tolerant Comput. Symp. (FTCS-14), Digest of Papers*, pp. 212-218, June 1984.

[309] L. Shen and S. Y. H. Su, "A Functional Testing Method for Microprocessors,"*IEEE Trans. Computers*, Vol. 37, pp. 1288-1293, October 1988.

[310] M. Shepherd and D. Rodgers, "Asynchronous FIFO's Require Special Attention," *Proc. Int. Test Conf.*, Philadelphia, PA, November 1985, pp. 445-450.

[311] S. D. Sherlekar and P. S. Subramanian, "Conditionally Robust Two-Pattern Tests and CMOS Design for Testability,"*IEEE Trans. CAD*, Vol. 7, pp. 325-332, March 1988.

[312] H. C. Shih and J. A. Abraham, "Transistor-Level Test Generation for Physical Failures in CMOS Circuits," *23rd Des. Aut. Conf.*, June 1986, pp. 243-249.

[313] T. Shimono, K. Oozeki, M. Takahashi, M. Kawai, and S. Funatsu, "An AC/DC Test Generation System for Gate Array LSIs," *Proc. Int. Test Conf.*, Philadelphia, PA, November 1985, pp. 329-333.

[314] M. Shirley, P. Wu, R. Davis, and G. Robinson, "A Synergistic Combination of Test Generation and Design for Testability," *Proc. Int. Test Conf.*, Washington, DC, September 1987, pp. 701-711.

[315] S. Shteingart, A. W. Nagle, and J. Grason, "RTG: Automatic Register Level Test Generator," *Proc. 22nd Des. Auto. Conf.*, Las Vegas, Nevada, June 1985, pp. 803-807.

[316] F. Siavoshi, "WTPGA: A Novel Weighted Test Pattern Generation Approach for VLSI Built-In Self-Test," *Proc. Int. Test Conf.*, Washington, DC, September 1988, pp. 256-262.

[317] G. M. Silberman and I. Spillinger, "Test Generation Using Functional Fault Simulation and the Difference Fault Model," *Proc. Int. Test Conf.*, Washington, DC, September 1987, pp. 400-407.

[318] G. M. Silberman and I. Spillinger, "An Approach to Test Generation of VLSI Designs using Their Functional Level Description," *Proc. Int. Conf. Comp. Des. (ICCD-87)*, Rye Brook, NY, October 1987, pp. 52-55.

[319] G. M. Silberman and I. Spillinger, "RIDDLE: A Foundation for Test Generation on a High Level Design Description," *Fault-Tolerant Computing Symposium (FTCS-18) Digest of Papers*, pp. 76-81, June 1988.

[320] G. M. Silberman and I. Spillinger, "G-RIDDLE: A Formal Analysis of Logic Designs Conducive to the Acceleration of Backtracking," *Proc. Int. Test Conf.*, Washington, DC, September 1988, pp. 764-772.

[321] N. Singh, *An Artificial Intelligence Approach to Test Generation*, Kluwer Academic Publishers, Norwell, MA, 1987.

[322] T. J. Snethen, "Simulator-Oriented Fault Test Generator," *Proc. 14th Des. Auto. Conf.*, 1977, pp. 88-93.

[323] F. Somenzi, S. Gai, M. Mezzalama, and P. Prinetto, "A New Integrated System for PLA Testing and Verification," *Proc. 20th Des. Auto. Conf.*, Miami Beach, FL, June 1983, pp. 57-63.

[324] F. Somenzi, S. Gai, M. Mezzalama, and P. Prinetto, "PART: Programmable Array Testing Based on a PARTitioning Algorithm," *IEEE Trans. on CAD*, Vol. CAD-3, pp. 142-149, April 1984. Also *Fault-Tolerant Comp. Symp. (FTCS-13) Digest of Papers*, Milan, Italy, June 1983, pp. 430-433.

[325] K. Son, "Rule Based Testability Checker and Test Generator," *Proc. Int. Test Conf.*, Philadelphia, PA, November 1985, pp. 884-889.

[326] K. Son and J. Y. O. Fong, "Automatic Behavioral Test Generation," *Int. Test Conf. Digest of Papers*, Philadelphia, PA, November 1982, pp. 161-165.

[327] V. P. Srini, "Test Generation from MacPitts Designs," *Proc. Int. Conf. Comp. Des. (ICCD'83)*, October 1983, pp. 53-56.

[328] N. C. E. Srinivas, A. S. Wojcik, and Y. H. Levendel, "An Artificial Intelligence Based Implementation of the P-Algorithm for Test Generation," *Proc. International Test Conference*, Washington, D.C., Sept. 1986, pp. 732-739.

[329] I. Stamelos, M. Melgara, M. Paolini, S. Morpurgo, and C. Segre, "A Multi-Level Test Pattern Generation and Validation Environment," *Proc. Int. Test Conf.*, Washington, DC, September 1986, pp. 90-96.

[330] K. E. Stoffers, "Test Sets for Combinational Logic – The Edge tracing Approach," *IEEE Trans. Computers*, Vol. C-29, pp. 741-746, August

1980.

[331] D. S. Suk and S. M. Reddy, "Test Procedures for a Class of Pattern-Sensitive Faults in Semiconductor Random-Access Memories,"*IEEE Trans. Computers*, Vol. C-29, pp. 419-429, June 1980.

[332] D. S. Suk and S. M. Reddy, "A March Test for Functional Faults in Semiconductor Random Access Memories,"*IEEE Trans. Computers*, Vol. C-30, pp. 982-985, December 1981.

[333] A. K. Susskind, "Testing by Verifying Walsh Coefficients,"*IEEE Trans. Computers*, Vol. C-32, pp. 198-201, February 1983. Also *Fault-Tolerant Comp. Symp. (FTCS-11) Digest of Papers,* Portland, Maine, June 1981, pp. 206-208.

[334] D. Svanaes and E. J. Aas, "Test Generation through Logic Programming,"*Integration, the VLSI Journal*, Vol. 2, pp. 49-67, March 1984.

[335] S. A. Szygenda, "A Software System for Diagnostic Test Generation and Simulation of Large Digital Systems," *Proc. 1969 National Electronics Conference*, 1969, pp. 657-662.

[336] S. A. Szygenda, "TEGAS – A Diagnostic Test Generation And Simulation System for Digital Computers," *Third Hawaii International Conference On System for Digital Computers*, Hawaii, 1970, pp. 163-166.

[337] S. A. Szygenda, "TEGAS2 – Anatomy of a General Purpose Test Generation and Simulation System for Digital Logic," *Proc. 9th Des. Auto. Conf.*, 1972, pp. 116-127.

[338] S. A. Szygenda and R. J. Smith, "A Processor for Software Diagnosis and Simulation of Digital Systems," *Second Annual Houston Conference On Circuits, Systems and Computers*, April 1970.

[339] A. Takahara and T. Nanya, "A Higher Level Hardware Design Verification," *Proc. IEEE Int. Conf. Comp. Design (ICCD-88)*, Rye Brook, NY, October 1988, pp. 596-603.

[340] Y. Takamatsu and K. Kinoshita, "CONT: A Concurrent Test Generation Algorithm,"*Fault-Tolerant Computing Symp. (FTCS-17) Digest of Papers*, pp. 22-27, July 1987.

[341] D. T. Tang and L. S. Woo, "Exhaustive Test Pattern Generation with Constant Weight Vectors,"*IEEE Trans. Computers*, Vol. C-32, pp. 1145-1150, Dec. 1983.

[342] S. M. Thatte and J. A. Abraham, "Test Generation for

Microprocessors,"*IEEE Trans. Computers*, Vol. C-29, pp. 429-441, June 1980.

[343] P. Thevenod-Fosse and R. David, "Random Testing of the Data Processing Section of a Microprocessor,"*Fault-Tolerant Comp. Symp. (FTCS-11) Digest of Papers*, pp. 275-280, June 1981.

[344] P. Thevenod-Fosse and R. David, "Random Testing of the Control Section of a Microprocessor,"*Fault-Tolerant Comp. Symp. (FTCS-13) Digest of Papers*, pp. 366-373, June 1983.

[345] J. J. Thomas, "Automated Diagnostic Test Program for Digital Networks,"*Computer Design*, pp. 63-67, August 1971.

[346] S. B. Timiri, "Devicec Test Generation Using Hardware Modeling," in *VLSI Design: Proc. Second Int. Workshop on VLSI Design*, R. Apte, V. D. Agrawal, and A. Prabhakar, Eds., Tata McGraw-Hill, New Delhi, India, 1988, pp. 448-455.

[347] C. Timoc, F. Scott, K. Wickman, and L. Hess, "Adaptive Probabilistic Testing of a Microprocessor," *Proc. Int. Conf. CAD (ICCAD-83)*, Santa Clara, CA, September 1983, pp. 71-72.

[348] K. E. Torku and B. M. Huey, "Petry Net Based Search Directing Heuristics for Test Generation," *Proc. 20th Des. Auto. Conf.*, Miami Beach, FL, June 1983, pp. 323-330.

[349] E. Trischler, "Guided inconsistent Path Sensitization: Method and Experimental Results," *Proc. Int. Test Conf.*, Philadelphia, PA, November 1985, pp. 79-86.

[350] M. E. Turner, D. G. Leet, R. J. Prilik, and D. J. McLean, "Testing CMOS VLSI: Tools, Concepts, and Experimental Results," *Proc. Int. Test Conf.*, Philadelphia, PA, November 1985, pp. 322-328.

[351] A. Tzidon, I. Berger, and M. Yoeli, "A Practical Approach to Fault Detection in Combinational Networks,"*IEEE Trans. Computers*, Vol. C-27, pp. 968-971, Oct. 1978.

[352] R. Ubar, "Test Pattern Generation for Digital Systems on the Vector Alternative Graph Model,"*13th Int. Symp. Fault Tolerant Computing (FTCS-13) Digest of Papers*, pp. 374-377, June 1983.

[353] J. G. Udell Jr., "Test Set Generation for Pseudo-Exhaustive BIST," *Proc. Int. Conf. CAD (ICCAD-86)*, Santa Clara, CA, November 1986, pp. 52-55.

[354] P. Varma and Y. Tohma, "PROTEAN – A Knowledge Based Test

Generator,'' *Proc. Custom Integrated Circ. Conf.*, Portland, OR, May 1987, pp. 78-81.

[355] P. K. Varshney, C. R. P. Hartman, and J. M. De Faria Jr., ''Application of Information Theory to Sequential Fault Diagnosis,''*IEEE Trans. Computers*, Vol. C-31, pp. 164-170, February 1982.

[356] S. Varszegi, ''An Interactive Method for the Generation of Tests Detecting Faults in Logic Netwroks,''*Int. J. Electronics*, Vol. 64, pp. 239-253, February 1988.

[357] R. Velazco, H. Ziade, and E. Kolokithas, ''A Microprocessor Test Approach Allowing Fault Localisation,'' *Proc. Int. Test Conf.*, Philadelphia, PA, November 1985, pp. 737-743.

[358] C. S. Venkatraman and K. K. Saluja, ''Transition Count Testing of Sequential Machines,''*Fault-Tolerant Comp. Symp. (FTCS-10) Digest of Papers*, pp. 167-172, October 1980.

[359] A. R. Virupakshia and V. C. V. P. Reddy, ''A Simple Random Test Procedure for Detection of Single Intermittent Fault in Combinational Circuits,''*IEEE Trans. Computers*, Vol. C-32, pp. 594-597, Jun. 1983.

[360] J. A. Waicukauski and E. Lindbloom, ''Fault Detection Effectiveness of Weighted Random Patterns,'' *Proc. Int. Test Conf.*, Washington, DC, September 1988, pp. 245-255.

[361] K. Walczak and E. Sapiecha, ''Multiple Fault Detection and Location in Large Combinational Circuits,''*Fault-Tolerant Comput. Symp. (FTCS-14), Digest of Papers*, pp. 134-140, June 1984.

[362] M. R. Wallace, ''A Tool Box Approach to Test Generation,'' in *VLSI Design: Proc. Second Int. Workshop on VLSI Design*, R. Apte, V. D. Agrawal, and A. Prabhakar, Eds., Tata McGraw-Hill, New Delhi, India, 1988, pp. 241-249.

[363] L. Y. Wei and J. Wei, ''Comments on Detection of Faults in Programmable Logic Arrays,''*IEEE Trans. Computers*, Vol. C-35, pp. 930-931, October 1986.

[364] R. S. Wei and A. Sangiovanni-Vincentelli, ''PLATYPUS: A PLA Test Pattern Generation Tool,''*IEEE Trans. on CAD*, Vol. CAD-5, pp. 633-644, October 1986.

[365] R. S. Wei and A. L. Sangiovanni-Vincentelli, ''New Front-End and Line Justification Algorithm for Automatic Test Generation,'' *Proc. Int. Test Conf.*, Washington, DC, September 1986, pp. 121-128.

[366] C. L. Wey and S. M. Chang, ''Test Generation of C-Testable Array Dividers,'' *Proc. IEEE Int. Conf. Comp. Design (ICCD-88)*, Rye Brook, NY, October 1988, pp. 140-144.

[367] D. J. Wharton, ''The HITEST Test Generation System – Overview,'' *Proc. Int. Test Conf.*, Philadelphia, PA, October 1983, pp. 302-310.

[368] T. W. Williams, Ed., *VLSI Testing*, North-Holland, Amsterdam, The Netherlands, 1986.

[369] W. Williams, ''An Automatic Test Generator for Programmable Logic Devices,'' *Proc. Int. Test Conf.*, Washington, DC, September 1987, pp. 658-667.

[370] S. Winegarden and D. Pannell, ''Paragons for Memory Test,'' *Int. Test Conf. Digest of Papers*, Philadelphia, PA, October 1981, pp. 44-48.

[371] H. J. Wunderlich, ''On Computing Optimized Input Probabilities for Random Tests,'' *Proc. 24th Des. Auto. Conf.*, Miami Beach, FL, June 1987, pp. 392-398.

[372] H. J. Wunderlich, ''Multiple Distributions for Biased Random Test Patterns,'' *Proc. Int. Test Conf.*, Washington, DC, September 1988, pp. 236-244.

[373] S. Xu and S. Y. H. Su, ''A Systematic Technique for Detecting and Locating Bridging and Stuck-At Faults in I/O Pins of LSI/VLSI Chips,''*Comput. Math. Applic.*, Vol. 13, No. 5/6, pp. 461-474, 1987.

[374] X. Xu and E. J. McCluskey, ''Test Generation and Fault Diagnosis for Multiple Faults in Combinational Circuits,''*13th Int. Symp. Fault Tolerant Computing (FTCS-13) Digest of Papers*, pp. 110-115, June 1983.

[375] C. L. Yang and G. M. Masson, ''A Distributed Algorithm for Fault Diagnosis in Systems with Soft Failures,''*IEEE Trans. Computers*, Vol. 37, pp. 1476-1480, November 1988.

[376] C. W. Yau, ''Concurrent Test Generation Using AI Techniques,'' *Proc. Int. Test Conf.*, Washington, D.C., September 1986, pp. 722-731.

[377] Z. Zhi-min and C. Ting-huai, ''A Near Optimal Algorithm for Test Generation in Sequential Circuit – Star Algorithm,''*Fault-Tolerant Comp. Symp. (FTCS-11) Digest of Papers*, pp. 233-237, June 1981.

SUBJECT INDEX

9 780792 390251